T0342427

GUIDELINES FOR
PROCESS SAFETY METRICS

GUIDELINES FOR PROCESS SAFETY METRICS

Center for Chemical Process Safety
New York, New York

A Joint Publication of the Center for Chemical Process Safety of the American Institute of Chemical Engineers and John Wiley & Sons, Inc.

Published by John Wiley & Sons, Inc., Hoboken, New Jersey.
Published simultaneously in Canada.

For general information on our other products and services or for technical support, please contact our Customer Care Department within the United States at (800) 762-2974, outside the United States at (317) 572-3993 or fax (317) 572-4002.

Wiley also publishes its books in a variety of electronic formats. Some content that appears in print may not be available in electronic format. For information about Wiley products, visit our web site at www.wiley.com.

Library of Congress Cataloging-in-Publication Data is available.

Guidelines for process safety metrics / Center for Chemical Process Safety.
    p. cm.
  Includes index.
  ISBN 978-0-470-57212-2 (cloth/cd)
  1. Industrial safety—Evaluation—Statistical methods. 2. Industrial management—Evaluation—Statistical methods. I. American Institute of Chemical Engineers. Center for Chemical Process Safety.
  T55.3.S72G85 2010
  660'.2804—dc22   2009031381

10 9 8 7 6

This book is one in a series of process safety guidelines and concept books published by the Center for Chemical Process Safety (CCPS). Please go to *www.wiley.com/go/ccps* for a full list of titles in this series.

CONTENTS

ITEMS ON THE FTP SITE ACCOMPANYING THIS BOOK

- CCPS "Process Safety Leading & Lagging Indicators," December 2007

- *Safety Performance Indicators Guidance for Industry and Safety Performance Indicators Guidance for Public Authorities and Communities*, ©OECD 2008, *available online at www.oecd.org/ehs* (reprinted with permission of OECD)

- HSG 254 *Developing process safety indicators. A step-by-step guide for chemical and major hazard industries* (ISBN 0717661806). © Crown copyright material is reproduced with the permission of the Controller of HMSO and Queen's Printer for Scotland.

- D. Cummings, *The Evolution and Current Status of Process Safety Metrics in DuPont* (presented at AIChE Spring National Meeting, April 2008)

- A. Hopkins, *Thinking About Process Safety Indicators* (reprinted from Safety Science 47 (2009) 460–65 with permission from Elsevier)

- F. Henselwood, *Use of Pareto Shape Parameter as a Leading Indicator of Process Safety Performance*, Process Safety Progress, February 26, 2009 (reprinted with permission of John Wiley & Sons, Inc.)

- K. Harrington, H. Thomas, S. Kadri, *Using Measured Performance as a Process Safety Leading Indicator* (Presented at the Global Congress on Process Safety, April 2008)

- *Examples of Safety Metrics Dashboards*

- *Full Listing of Potential Process Safety Metrics to Consider (Based on Risk Based Process Safety Elements)*

The link for the ftp site is:

ftp://ftp.wiley.com/public/sci_tech_med/gl_metrics

ACRONYMS AND ABBREVIATIONS

ACC	American Chemistry Council
AIChE	American Institute of Chemical Engineers
ANI	American Nuclear Insurers
API	American Petroleum Institute
CAPP	Chemical Accident Prevention Program
CCPS	Center for Chemical Process Safety
CFR	Code of Federal Regulations
COMAH	Control of Major Accident Hazards (U.K. HSE Regulation)
CSB	U.S. Chemical Safety and Hazard Investigation Board
EPA	U.S. Environmental Protection Agency
EFCE	European Federation of Chemical Engineering
EU	European Union
FMEA	Failure Modes and Effects Analysis
GEMI	Global Environmental Management Initiative
HIRA	Hazard Identification and Risk Analysis
HSE	Health and Safety Executive (U.K.)
ILO	International Labour Organization
INPO	Institute of Nuclear Power Operations
MI	Mechanical Integrity
MOC	Management of Change
NEP	National Emphasis Program
NGO	Non-Governmental Organization
OECD	Organization for Economic Cooperation and Development
OII	Occupational Injury and Illness
OSHA	U.S. Occupational Safety and Health Administration
P&ID	Piping and Instrumentation Diagram
PHA	Process Hazard Analysis
PSI	Process Safety Incident
PSIC	Process Safety Incidents Count
PSISR	Process Safety Incident Severity Rate
PSM	Process Safety Management (U.S. OSHA Regulation)

RAGAGEP	Recognized and Generally Accepted Good Engineering Practice
RBPS	Risk-Based Process Safety
RMP	Risk Management Program; Risk Management Plan
ROP	Reactor Oversight Program (U.S. Nuclear Regulatory Commission)
SAICM	Strategic Approach to International Chemicals Management
SAT	Systematic Approach to Training
TCPA	Toxic Catastrophe and Prevention Act
TQ	Threshold Quantity
UNEP	United Nations Environment Programme
VPP	Voluntary Protection Programs (U.S. Occupational Safety and Health Administration)

GLOSSARY

Asset Integrity
Work activities that help ensure that equipment is properly designed, is installed in accordance with specifications, and remains fit for purpose over its life cycle.

Audit
A systematic, independent review to verify conformance with prescribed standards of care using a well-defined review process to ensure consistency and to allow the auditor to reach defensible findings.

Controls
Engineered mechanisms and administrative policies/ procedures implemented to prevent or mitigate incidents.

Element
Basic division in a process safety management system that correlates to the type of work that must be done (e.g., management of change [MOC]).

Facility
The physical location or site where the management system activity is performed.

Hazard
Chemical or physical conditions that have the potential for causing harm to people, property, or the environment.

Hazard Identification and Risk Analysis (HIRA)
A collective term that encompasses all activities involved in identifying hazards and evaluating risk at facilities, throughout their life cycle, to make certain that risks to employees, the public, or the environment are consistently controlled within the organization's risk tolerance.

Incident
An unusual or unexpected event that either resulted in, or had the potential to result in, serious injury to personnel, significant damage to property, adverse environmental impact, or a major interruption of process operations.

Incident Investigation
A systematic approach for determining the causes of an incident and developing recommendations that address the causes to help prevent or mitigate future incidents.

Knowledge (or Process Safety Knowledge)	Knowledge is related to information, which is often associated with policies, and other rule-based facts. It includes work activities to gather, organize, maintain, and provide information to other process safety elements. Process safety knowledge primarily consists of written documents such as hazard information, process technology information, and equipment-specific information.
Lagging Metric	A retrospective set of metrics based on incidents that meet an established threshold of severity.
Leading Metric	A forward-looking set of metrics that indicate the performance of the key work processes, operating discipline, or layers of protection that prevent incidents.
Life Cycle	The stages that a physical process or a management system goes through as it proceeds from birth to death. These stages include conception, design, deployment, acquisition, operation, maintenance, decommissioning, and disposal.
Management of Change (MOC)	A system to identify, review, and approve all modifications to equipment, procedures, raw materials, and processing conditions, other than "replacement in kind," prior to implementation.
Management System	A formally established set of activities designed to produce specific results in a consistent manner on a sustainable basis.
Mechanical Integrity	A program that helps ensure that equipment is properly designed, installed in accordance with specifications, and remains fit for purpose over its life cycle. Also, asset integrity.
Near-Miss Incident	The description of less severe incidents (i.e., below the threshold for inclusion in a lagging metric), or unsafe conditions that activated one or more layers of protection. Although these events are actual events (i.e., a "lagging" metric), they are generally considered to be a good indicator of conditions that could ultimately lead to a severe incident.
OSHA Process Safety Management (OSHA PSM)	A U.S. regulatory standard that requires use of a 14-element management system to help prevent or mitigate the effects of catastrophic releases of chemicals or energy from processes covered by the regulations (49 C.F.R. §1910.119).
Procedures	Written step-by-step instructions and associated information (cautions, notes, warnings) that describe how to safely perform a task.

Process	A broad term that includes the equipment and technology needed for petrochemical production, including reactors, tanks, piping, boilers, cooling towers, refrigeration systems, etc.
Process Safety	A disciplined framework for managing the integrity of operating systems and processes handling hazardous substances by applying good design principles, engineering, and operating practices. It deals with the prevention and control of incidents that have the potential to release hazardous materials or energy. Such incidents can cause toxic effects, fire, or explosion and could ultimately result in serious injuries, property damage, lost production, and environmental impact.
Process Safety Culture	The combination of group values and behaviors that determines the manner in which process safety is managed. A sound process safety culture refers to attitudes and behaviors that support the goal of safer process operations.
Process Safety Management	The application of management systems to the identification, understanding, and control of process hazards to prevent process-related injuries and incidents; it is focused on prevention of, preparedness for, mitigation of, response to, and restoration from catastrophic releases of chemicals or energy from a process associated with a facility.
Process Safety Metric	A standard of measurement or indicator of process safety management efficiency or performance.
Reliability	The probability that an item is able to perform a required function under stated conditions for a stated period of time or for a stated demand.
Reliability Analysis	The determination of reliability of a process, system, or equipment.
Risk-Based Process Safety	The CCPS's process safety management system approach that uses risk-based strategies and implementation tactics that are commensurate with the risk-based need for process safety activities, availability of resources, and existing process safety culture to design, correct, and improve process safety management activities.
Safe Operating Limits	Limits established for critical process parameters, such as temperature, pressure, level, flow, or concentration, based on a combination of equipment design limits and the dynamics of the process.
Safeguard	See Controls.

Safety Instrumented System   The instrumentation, controls, and interlocks provided for safe operation of the process.

Six Sigma   A metric for measuring defects and improving quality. Also, a robust business improvement methodology that focuses an organization on customer requirements, process alignment, analytical rigor, and timely execution.

Stakeholder   Individuals or organizations that can (or believe they can) be affected by the facility's operations, or that are involved with assisting or monitoring facility operations.

Sustainability   Meeting the needs of the present without compromising the ability of future generations to meet their own needs.

Training   Practical instruction in job and task requirements and methods. Training may be provided in a classroom or at the workplace, and its objective is to enable workers to meet some minimum initial performance standards, to maintain their proficiency, or to qualify for promotion to a more demanding position.

Transparency   Openness of an organization with regard to sharing information about how it operates.

Triple Bottom Line   Expanding the traditional business-reporting framework to take into account environmental and social performance in addition to financial performance.

ACKNOWLEDGMENTS

The American Institute of Chemical Engineers (AIChE) and the Center for Chemical Process Safety (CCPS) express their appreciation and gratitude to all members of the Process Safety Metrics Subcommittee and their CCPS member companies for their generous support and technical contributions in the preparation of these *Guidelines*. The AIChE and CCPS also express their gratitude to the team of authors from AcuTech Consulting Group.

PROCESS SAFETY METRICS SUBCOMMITTEE MEMBERS:

Tim Overton, Chair	The Dow Chemical Company
Michael Broadribb	BP
Cho Nai Cheung	Contra Costa County Health Services Department
Elroy Christie	Honeywell
Susie Cowher	INEOS
Eric Freiburger	NOVA Chemical
Harry Glidden	DuPont
Kent Goddard	Solutia
Rick Griffin	Chevron Phillips Chemical Company
Karen Haase	American Petroleum Institute
Kenneth Harrington	Chevron Phillips Chemical Company
Steven Hedrick	Bayer Material Science
Shakeel Kadri	Air Products and Chemicals
Lisa Long	U.S. Occupational Safety and Health Administration
Jack McCavit	CCPS Emeritus
Mark Miner	Nalco
Jeff Philiph	Monsanto
Cathy Pincus	ExxonMobil
William Ralph	BP
Isador (Irv) Rosenthal	The Wharton School, University of Pennsylvania
Randall Sawyer	Contra Costa County Health Services Department
S. L. Sreedhar	Santos Ltd.
Tomaysa Sterling	American Chemistry Council

Lara Swett   National Petrochemical Refiners Association
Steven Tanzi   PriceWaterhouseCoopers
Brian Wilson   Rohm & Haas

CCPS Staff Consultant: Dan Sliva

CCPS wishes to acknowledge the many contributions of the AcuTech Consulting Group staff members who wrote this book especially the principal authors Dorothy Kellogg and Greg Keeports. The authors wish to thank the following AcuTech personnel for their technical contributions and review: Christie Arseneau, David Moore, and Lee Salamone.

Before publication, all CCPS books are subjected to a thorough peer review process. CCPS gratefully acknowledges the thoughtful comments and suggestions of the peer reviewers. Their work enhanced the accuracy and clarity of these guidelines.

Peer Reviewers:

Jonathan Carter   Marsh Marine and Energy Ltd.
William R. Corcoran   Nuclear Safety Review Concepts Corporation
Kenneth Daigle   BP
F. Russ Davis   Solutia
Tom Garvin   SAIC
Bill Green   NOVA Chemical
Dennis Hendershot   CCPS Emeritus
John Herber   3M Company
Andrew Hopkins   Australian National University
Christian Jochum   European Process Safety Centre
Bert Knegtering   Honeywell Safety Solutions
William E. Lash   BP
Ian McPherson   U.K. Petroleum Industry Association
James P. Miller   ConocoPhillips
John Murphy   CCPS Emeritus
Sheri Sammons   INEOS Olefins and Polymers USA
Angela Summers   SIS-TECH Solutions, LP
Ian Travers   U.K. Health and Safety Executive
Lee Valentine   BP

PREFACE

Process safety metrics has been an area of strong interest to industry, regulators, and experts as key enablers to improve process safety performance across industry. The expectation is that improved leading and lagging metrics and leadership engagement will lead to breakthroughs in performance, resulting in much lower rates of significant process safety incidents.

This interest led to the CCPS Technical Steering Committee authorizing the creation of a project committee to develop a book of guidelines for the development and use of leading and lagging process safety metrics. To achieve broader industry acceptance, CCPS invited representatives from many chemical and petroleum companies, trade associations, labor groups, regulators, and academics involved in the field of process safety, as well as other key stakeholders or subject mater experts to participate in this committee's activities.

The interest in metrics continues to increase due to the attention given by both the Baker Panel and the U.S. Chemical Safety Board's reports regarding the BP Texas City incident. The need for better metrics was highlighted as one of four key issues on the cover of the CSB report. The Baker Panel also identified this issue as one its top-10 recommendations and dedicated several pages of its final report to discussions of both leading and lagging process safety metrics. The Baker Panel recommended that "organizations should develop leading process safety performance indicators that, if monitored, can be used to limit or prevent process-related incidents."

As stated by Lord Kelvin (1824–1907), "If you cannot measure it, you cannot improve it." Metrics are needed to measure where improvements are occurring and where more improvement is needed (i.e., where intervention is needed). However, improvements will not occur simply from implementation and collection of metrics—the plant/business must *act* upon the metrics.

Corporate leadership must set the tone and expectation that metrics will be accurately reported and be engaged to ensure that evaluation of metrics and improvement occur. Establishing the right culture of expectation and intolerance for deviation or variability is essential to achieving good process safety performance in the chemical and allied processing industries. Implementing a comprehensive metric system is a vital element of establishing and maintaining such culture. Corporate and line management should remember that if a plant or plant leader is not managing process safety well, they are probably not managing other things well either.

However, the value of good process safety metrics is not just for corporate leaders. Having effective leading metrics can also be invaluable to front-line leaders who are being held accountable for results in all areas of performance by their leadership. Plant leaders are under continuous pressure to reduce costs, increase productivity, and maintain safety. Metrics are more readily available to demonstrate the impact of cost reduction or productivity measures. Without an equivalent quick-response process safety metric (leading metric) to demonstrate that safety programs may degrade if some cost-cutting measures are implemented, the plant leader is pressured to make changes and simply hope there is no negative impact upon safety. Good leading metrics and management of change procedures can be used to detect whether cost reduction measures have affected or will affect safety, or where improvements in the safety program are needed.

The purpose of this book is to provide guidance to many levels of the organization when implementing or improving existing corporate process safety metrics. Although the process safety leaders in the company will have the strongest interest, it is equally important that others in leadership roles also read this book and work together with the process safety leaders in selecting and implementing the appropriate metric programs.

The overall "tone" from each leader in the organization is necessary to drive improvement in the right areas to deliver results. The use of leading metrics will assist the organization in being forward-looking rather than simply analyzing what went wrong afterwards. Management's commitment to act in response to identified problems is as important, if not more important, than the metrics themselves.

The American Institute of Chemical Engineers (AIChE) has been closely involved with process safety and loss control issues in the chemical and allied industries for more than four decades. AIChE publications and symposia have become information resources for those devoted to process safety and environmental protection.

AIChE created the Center for Chemical Process Safety (CCPS) in 1985 after the chemical disasters in Mexico City, Mexico, and Bhopal, India. The CCPS is chartered to develop and disseminate technical information for use in the prevention of major chemical accidents. The center is supported by more than 100 chemical process industry sponsors that provide the necessary funding and professional guidance to its technical committees. The major product of CCPS activities has been a series of guidelines to assist those implementing various elements of a process safety and risk management system. This book is part of that series.

CCPS strongly encourages companies around the globe to adopt and implement the recommendations contained within this book.

1

INTRODUCTION

1.1   AN INTRODUCTION TO PROCESS SAFETY AND METRICS

Process safety management[1] is a disciplined framework for managing the integrity of operating systems and processes handling hazardous substances by applying good engineering, operating, and maintenance practices. It deals with the prevention and control of risks that have the potential to release hazardous materials or energy. Such incidents can result in a toxic release, fire, or explosion and could ultimately result in serious injuries, property damage, environmental degradation, and lost production.

Process safety incidents are rarely caused by a single catastrophic failure, but rather by multiple events or failures that coincide and collectively result in an incident. This relationship between simultaneous or sequential failures of multiple systems is illustrated by the "Swiss cheese model,"[2] as shown in Figure 1.1, where hazards are contained by multiple protective barriers that may have weaknesses or "holes."

The barriers (represented as individual slices of "cheese") are elements of a process safety management system or some other layer of protection designed to prevent an incident from occurring. The holes in the slice of cheese represent deficiencies (or failures) in those barriers, gaps that may allow an event to escalate into an incident. When holes align, hazardous energy or chemical may be released, resulting in the potential for harm. Barriers may be physically engineered containment or behavioral controls dependent on people. Holes may be latent, incipient, or actively opened by people.

---

[1] "Process safety" is the term used in the process industries for what, in a more general technological context, is called "system safety."

[2] The Swiss cheese model of accident causation was originally proposed by British psychologist James T. Reason and has since gained widespread acceptance in many risk-analysis and management fields including process safety.

Figure 1.1 Swiss Cheese Process Safety Model (CCPS, 2007b)

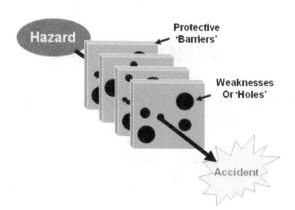

This analogy can also describe two types of metrics—leading and lagging. Events that occur by passing through gaps in a process safety management system or protective system (e.g., loss of primary containment events, near misses, or process safety incidents) can be described as "lagging indicators." However, detecting and measuring a failure of a management system or protective system (i.e., a hole in the "cheese") before an incident occurs (e.g., failure to complete a scheduled mechanical integrity system, failure to use an appropriate management of change (MOC) process) are described as "leading metrics."

Serious incidents may be predicated by a number of less-severe related incidents resulting in minor or even no loss. Such predicating events may be low-consequence incidents such as loss of containment into a diked area, near misses, or failures in which no injuries, damage, or loss occurred. This relationship between no- or low-impact events and actual process safety incidents is demonstrated in the process safety pyramid (see Figure 1.2).[3]

In *Guidelines for Risk Based Process Safety,* CCPS defines process safety management as a "management system that is focused on prevention of, preparedness for, mitigation of, response to, and restoration from catastrophic releases of chemicals or energy from a process associated with a facility" (CCPS, 2007a). The process safety management system comprises the design, procedures, and hardware needed to operate and maintain the process safely throughout the operational life cycle (CCPS, 1999). Such a process safety management system will include metrics to identify and measure not only actual process safety incidents that meet an established reporting threshold, as well as metrics to identify

---

[3] Use of the process safety pyramid in selecting metrics is discussed in Chapters 3 and 6.

lower-severity incidents, near misses (no-loss incidents), and unsafe behaviors. Process safety metrics should track performance of individual system components to ensure that the process safety systems are performing as intended and to identify nonconformities within systems before they can cascade and result in a serious, reportable incident.

Process safety metrics provide a means to measure activity, status, or performance against requirements and goals. Monitoring and analyzing such performance enables organizations to identify and track not only current performance but also trends, both improvements and degradations, so that corrective actions are taken as needed. An organization that expects and maintains performance within operating specifications and that monitors activities or behaviors critical to overall safety operations (e.g., training, management of change) is more likely to avoid major failures, including catastrophic events. In the event of an incident, process safety metrics can provide critical information to identify contributing and root causes of the failures and a system to track subsequent system improvements. The need to monitor behavior includes the behavior of engineers and managers, not just operators. Unsafe behavior includes things like the unit manager deciding that he can delay a required safety inspection to save money on his maintenance budget or to avoid a plant shutdown, or an engineer deciding that he does not need to do a management of change review of a piping modification.

1.2   PURPOSE OF THIS BOOK

This book has many purposes, including the following:

- To provide guidelines and examples of effective practices for the development and use of process safety leading and lagging metrics;
- To provide the reader with summaries and references to other useful resources;
- To convey basic information about performance indicators such as the what, when, where, why, and to whom they are useful;
- To provide guidance to companies on how to collect, evaluate, and communicate process safety metrics;
- To provide guidance to companies at corporate and site levels on how to use performance metrics effectively to improve process safety performance;
- To provide sufficient examples such that readers gain an understanding of how performance metrics can be successfully applied over the short and long term; and
- To encourage the adoption of a set of consensus process safety metrics comparable to occupational safety metrics.

Figure 1.2 Process Safety Pyramid (CCPS, 2007b)

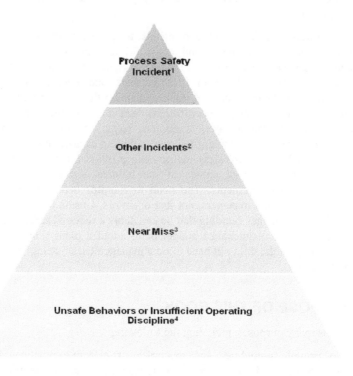

1. Incidents which meet the threshold of severity which should be
    reported as part of the process safety metric.
2. Incidents which didn't meet the definition of PS incident metric (e.g., all
    other Loss of Primary Containment or fire).
3. System failures which could have led to an incident (e.g., instrument
    had failed, pipe wall thickness is low).
4. Other Process Safety Factors, e.g., Equipment Selection, Engineering
    Design, Specification of Inspection Frequency and Technique, Unsafe
    Behaviors and Insufficient Operating Discipline.

1.3   KEY AUDIENCES FOR THE GUIDELINES

This book is primarily written for those within the chemical process and allied industries responsible for process safety including:

- *Process Safety Specialists*—those at the corporate and facility levels responsible for the process safety system including tailoring the system to specific facilities and using metrics to monitor and maintain or improve process safety performance.

- *Line Management (area managers, unit managers)*—those responsible for collecting metrics data and ensuring that the behaviors in the work area are consistent with the expectations of the process safety system.

- *Site Management (site managers, plant managers, facility managers, health safety, and environmental (HSE) managers)*—those who set site performance goals based on management expectations and use metrics data to measure and improve the process safety performance at a given site.

- *Corporate Leaders*—those providing the leadership commitment, setting the expectations, for a process safety performance and the supporting metrics system including the resources necessary to develop and implement such a program throughout the corporation to improve process safety.

Other audiences and stakeholders interested in improving process safety are also likely to find this book useful and informative and may include:

- Production personnel (operators), including representatives of labor, who are concerned with the safety of the operators

- Those within a company who wish to make comparisons between sites with process safety programs

- Personnel responsible for mechanical integrity (MI), management of change (MOC), and other elements of the process safety system who can use metrics to measure the performance of their systems;

- Corporate communicators responsible for communicating the company's safety performance within the company and to the public

- Industry trade groups that want to promote the safety of the industries they represent and show progress towards sustainable process safety improvement;

- Government regulators and analysts who are concerned with regulating or analyzing companies and industries, including those in the fields of public health, emergency response, and local land-use planning

- Non-governmental organizations (NGOs) addressing environmental, public safety, and other chemical-related issues

- Members of the public, especially those near a process industry facility, who could be affected in the event of a major or catastrophic accident

The scope of this book is intended to be globally applicable, and the concepts and approaches will be useful to all companies seeking to continuously improve their process safety performance. However, many of the references and examples are based on U.S. conditions and programs.

1.4   AN ORGANIZATION'S PERSONNEL HIERARCHY

Throughout these guidelines, references will be made to an organizational personnel hierarchy—the levels of the organization generally spanning operating and maintenance personnel to the board of directors—each have different responsibilities, needs, and interests with respect to process safety metrics. Different organizations may have different terms for different personnel levels, but for the purpose of the book the following descriptions will be used:

- *Operating and Maintenance Personnel (operators, mechanics, craft personnel)*—individual contributors who carry out specific tasks and/or procedures

- *Supervisors (foremen, first-line leaders)*—individuals who provide direct supervision for the operating, maintenance, and craft personnel – could be seen as first level of line management

- *Line Management (area managers, unit managers)*—individuals who have management responsibility for organizational units of process operating facilities and maintenance organizations, normally situated in a single plant site or operating area

- *Site Management (site managers, plant managers, facility managers)*—individuals who have management responsibility for an entire site, usually based upon geographical boundaries, and may include multiple production units on the site

- *Business Leaders*—individuals who have management responsibility for a business unit or portion of a company, including oversight for a manufacturing plant(s) associated with business; budget approval for operations; and oversight of marketing, sales, and other functional areas that generate the production demand

- *Corporate Leaders*—individuals who have responsibility at the highest levels of a company for business and/or company support functions; focus upon the strategic issues and planning for the enterprise; often have overall management responsibility for a business or groups of businesses; and usually do not have day-to-day operational responsibility

- *Board of Directors*—body of appointed persons, including company officers and outside directors, who jointly oversee the activities of an enterprise including ensuring that the enterprise is managing risks properly

1.5 ORGANIZATION OF THIS GUIDELINE

This book is organized in a logical flow for developing and implementing a process safety metrics program. Because some readers may read the book from beginning to end while others may select specific chapters, the chapters have been written to stand alone as well as part of the book as a whole.

- Chapter 2—*Why Implement Process Safety Metrics,* especially the role of metrics in management
- Chapter 3—*Process Safety Management Metrics* commonly used in process safety management systems including leading, lagging, and near miss; and activity and outcome, external and internal metrics as well as characteristics of successful metrics
- Chapter 4—*Choosing Appropriate Metrics* based on identified process safety goals and objectives
- Chapter 5—*Implementing a Metrics Program,* including implementation strategy and framework, analyses, and program rollout
- Chapter 6—*Communicating Results* to various internal and external audiences and tools for reaching those audiences
- Chapter 7—*Using Metrics to Drive Performance Improvements* through identification of system strengths and weaknesses, holding responsible parties accountable, engaging the public, conducting management reviews, cultivating a positive process safety culture, and communicating successes
- Chapter 8—*Improving Industry-Wide Performance* through the adoption of common process safety metrics and definitions within companies and industry sectors, and across the processing industries
- Chapter 9—*Future Trends in the Development and Use of Process Safety Metrics* for improving process safety performance and broader societal interests

The guidelines include examples of metrics and how to establish, use, and update them. These examples are garnered from several sources including the recently published CCPS "Process Safety Leading and Lagging Metrics" brochure, which is included on the accompanying CD, and the CCPS *Guidelines for Risk Based Process Safety.* Every attempt is made to be consistent with other published guidelines from CCPS while reflecting growth and change in the development and use of process safety metrics.

1.6 USING THIS GUIDELINE

It is hoped that process safety performance throughout the process industries will improve as an increasing number of companies adopt more extensive and more rigorous process safety metrics as part of their process safety management systems. Companies and facilities that do this should ultimately see improvement in their process safety performance, which improves the performance of the sector more broadly. In addition, benchmarking and other comparisons can be made within and across sectors when companies adopt common public-facing metrics. Such public comparisons both acknowledge high performers and provide improvement incentives for all.

REFERENCES

Center for Chemical Process Safety, *Guidelines for Process Safety in Batch Reaction Systems*, American Institute of Chemical Engineers, New York, 1999
Center for Chemical Process Safety, "Process Safety Leading and Lagging Metrics," American Institute of Chemical Engineers, New York, 2007
Reason, J., *The contribution of latent human failures to the breakdown of complex systems*, Philosophical Transactions of the Royal Society (London), series B. 327: 475–84 (1990)

2

WHY IMPLEMENT PROCESS SAFETY METRICS

In January 1997 an explosion and fire on the Hydrocracker Unit at the Tosco Avon oil refinery in Martinez, California, resulted in one death and 46 injured workers of which 13 were hospitalized. There had been a significant number of temperature excursions that had taken place prior to the 1997 incident with no post-incident investigation and remedial actions pushed through to full implementation.

Reports on the incident identified deficiencies in the safety management oversight system including weaknesses in programs to:

- Identify and control hazards
- Establish and maintain safe and achievable operating limits
- Track, investigate, and address near misses
- Conduct and document management of change reviews

A metrics program that identified and implemented appropriate metrics would allow the facility to track whether PHAs and MOCs are completed prior to work or a change and whether near misses had been documented and investigated.

Source: *Investigation Report: Refinery Fire Incident, Tosco Avon Refinery,* U.S. Chemical Safety Board, 2001

Achieving high-quality process safety management is critical to any enterprise that operates hazardous processes. Measuring process safety systems performance is a key to achieving and maintaining superior performance. Measuring performance to improve execution of work systems is a long-proven and successful technique used throughout industry. Activities in industry—including those in the process industries—can be described as work processes and procedures that are overseen by management systems. These systems share the common characteristic that performance of the detailed work processes is monitored and evaluated by tracking key

The key process safety objective is to identify failures, gaps or conditions and to correct them before they contribute to a major process safety incident.

performance indicators. This is true from back-office activities to the operating floor in the processing facilities.

Properly selected metrics that fit with the detailed objectives of an organization will identify the successes and point out the weaknesses of the system. The stakes for the process industries to sustain strong process safety performance are quite high given the potential outcomes from catastrophic process safety incidents. Using proven performance monitoring methodologies to sustain and improve reliable execution for process safety management systems has a huge potential return.

The need for metrics is particularly important in process safety, in no small measure because the hazards may not be readily evident. Unlike some other safety risks where dangerous situations are more apparent—such as unsafe scaffolding, unsecured cables, trenches, and other excavations—information on the status and safety of hazardous containment systems (e.g., internal corrosion, an improperly sized relief valve) is not generally visible. Without a constant and reliable flow of information on process safety performance and management systems, leaders may, in essence, be flying blind.

The authors of the *Guidelines for Risk Based Process Safety* (RBPS) (CCPS, 2007a) described the need for constant vigilance as the price of maintaining an effective process safety management system. An operator not only must be vigilant (aware of both past and current performance), but must not assume that current performance will be maintained, much less improved, without intentional evaluation of critical parts of systems and their performance. Performance measurement and metrics are a critical part of the RBPS system.

These concepts advocate good process safety management as a primary means to prevent process safety incidents. Just because a catastrophic incident has not occurred in the past does not mean that such an incident could not happen in the future. No one comes to work knowing that this is the day an incident will occur. No one can predict when multiple factors will align to cause an incident, and likewise, no one can know the magnitude of a specific process safety incident in advance even though predictive modeling may have been done. The difference between a catastrophic incident and a less-severe impact may be a matter of good fortune, such as time of day or wind direction, since consequences most often fall short of an incident's potential maximum impact.

Knowledgeable practitioners consider implementing and sustaining reliable process safety management practices as the most effective approach to avoiding process safety incidents. Even though the timing of process safety incidents cannot be predicted, a plant that continues on a path of poor process safety management performance will have an increased likelihood of incidents. Performance measurement combined with proper corrective and improvement actions for process safety provide confidence that process safety incidents will be rare events and progress toward the goal of eliminating them is achievable.

2.1 PREVENTING PROCESS SAFETY INCIDENTS

The primary goal of a process safety management system is to eliminate process safety incidents and potential catastrophic impacts upon people, the environment, equipment and production. Process safety is dedicated to safe process design and reliable process operations to avoid failures in processing and related equipment that can lead to a process safety incident. The process safety management system comprises the design, procedures, and hardware needed to operate and maintain a process safely throughout its operational life cycle (CCPS, 1999). For example, *Guidelines for Risk Based Process Safety* identified 20 process safety elements covering an organization's commitment to process safety, understanding of hazards, management of risk, and learning from experience (CCPS, 2007a). The CCPS also published *Inherently Safer Chemical Processes: A Life Cycle Approach* and *Guidelines for Engineering Design for Process Safety*, both of which provide detailed concepts that can form a basis for tracking performance. Process safety professionals recognize that weaknesses in any of the defined process safety elements can contribute to a process safety incident. It is important to ensure reliable implementation and execution of a broad set of process safety–related activities, and a performance-monitoring system that tracks performance of these activities is necessary to meet process safety goals and objectives.

> *Past history is no indication of future performance and returns are not guaranteed. Past good performance is no guarantee that incidents will not occur in the future.*

Process safety incidents can lead to disruption of production operations that may last hours, days, weeks, or even years. *Incidents that Define Process Safety* is replete with examples of fires, explosions, and releases that have had a significant impact on the chemical and refining industries' approaches to modern process safety (CCPS, 2008a). Severe incidents can lead to significant loss of life, destruction of property, and damage to the environment and surrounding infrastructure as well as regulatory and financial costs, civil and criminal liabilities, and even threats to the enterprise's survival (CCPS, 2006). For example:

- The economic impact of the December 2005 vapor cloud explosion at the Buncefield Oil Storage Terminal (Hertfordshire, UK) has been estimated at up to £1 billion ($1.8 billion 2005 dollars, €1.5 billion 2005 Euros) including compensation for loss, costs to the aviation sector, the emergency response, and the cost of the investigation. Local office buildings, schools, churches, and other structures were significantly damaged or even destroyed and, because the terminal supplied 30 percent of its fuel, Heathrow Airport had to ration fuel for several months following the explosion (Buncefield, 2008).

- BP has set aside $2.125 million to settle civil claims resulting from the 2005 explosion and fire at its Texas City refinery; this is in addition to a

$21.4 million OSHA penalty and a $50 million criminal fine and three years' probation (BP, 2007).

- In 2000, a federal grand jury indicted the president of Concept Sciences, Inc., for alleged criminal violations of the OSHA process safety management (PSM) standard; he faced a maximum 24 months in prison, a $3 million fine, and one year of probation[4] (CW, 1999).

- In 2002, a French court found the president of Total guilty of manslaughter in the death of six operators from the fluid catalytic cracking unit explosion at the La Mede refinery. He was sentenced to 18 months in prison (suspended) and fined €4,500 ($4,000) (CCPS, 2008a).

- The European Union enacted the directive on the control of major-accident hazards involving dangerous substances in response to the 1976 dioxin release from the ICMESA facility near Seveso, Italy; the United States enacted a similar Emergency Preparedness and Community Right-to-Know Act in response to the 1984 Bhopal disaster and 1989 Phillips Pasadena, Texas, refinery explosion and fire.

- The 1984 release of methyl isocyanate from the Union Carbide plant in Bhopal, India, not only resulted in shutting down that facility and Union Carbide's ultimate cessation of all operations in India, but also negatively affected the company's stock price in the short term and led to a hostile takeover attempt by a competitor (which UCC was able to thwart).

Performance of the process industry must serve the government and society as a whole. Regulatory agencies across the world are becoming more knowledgeable of the hazards and risks associated with the process industries. Every process safety incident that injures people or threatens the public is scrutinized thoroughly. If regulators and legislators become concerned with poor process safety performance, as evidenced by incident rates, they will develop policies and programs to address the problem. For example, in 2007 the U.S. OSHA established the National Emphasis Program (NEP) for refineries after OSHA determined that the number of incidents in the refining sector exceeded those in other sectors. Eliminating process safety incidents is the most attractive option for long-term operational viability. Establishing the right culture of expectations and intolerance of deviation or variability includes implementing a comprehensive metrics system. Implementing a process safety metrics system—especially one accepted and widely adopted within the industry—is a key factor in achieving this goal.

---

[4] The court subsequently overturned the criminal indictment, ruling that the OSHA PSM regulation was so ambiguous a reasonable person could not have determined whether it applied to the defendant's activities and that the Administrative Procedures Act prohibits application of OSHA's informal interpretations in a criminal case (Surrick, Judge R. Barclay, *United States of America v. Irl "Chip" Ward*, U.S. District Court for the Eastern District of Pennsylvania, Sept. 5, 2001).

2.2 BENEFITS FROM MEASURING PERFORMANCE

Long-standing consensus among business management experts indicates that measuring performance of business processes is critical to determining performance strengths and weaknesses. Achieving reliable performance of operating systems will lead to superior results. The successful use of performance improvement techniques based upon this philosophy has generated improved process safety results. Companies continue to strive for greater efficiency, reliability, and cost- effectiveness to remain competitive. At the same time, they must maintain a high level of performance, such as providing good and reliable product quality. In addition, many managers and executives are accustomed to and comfortable with interpreting and relying on numerical data. Properly defined and understood metrics can give management confidence that the right things are being managed and tracked. There are lessons from the quality efforts adopted for business processes and manufacturing operations:

- How to monitor performance
- How to determine causes for unwanted outcomes
- How to develop strategies to improve performance

The concept of continuous improvement is relatively simple, and it is often stated as the Plan-Do-Check-Act cycle. The basic strategy follows these steps:

1. Define what is to be accomplished (establish objectives).

2. Develop and execute the improvement plan.

3. Track what is done and how well it is done.

4. Identify opportunities to improve the plan and/or execution and make appropriate changes.

5. Continue the Plan-Do-Check-Act cycle until the acceptable level of performance is attained and the desired performance is maintained.

Organizations' quality efforts describe various methodologies to implement a continuous improvement program. A common link to these quality programs and process safety management is the reliable execution of work processes, based upon specific procedures that are necessary to carry out defined tasks. Within the process safety arena multiple elements must be performed in a highly reliable manner to avoid process safety incidents. To manage this array of elements reliably requires appropriately designed performance tracking systems.

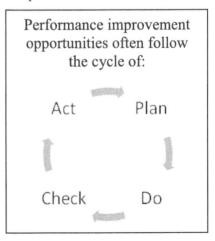

Performance improvement opportunities often follow the cycle of:

Act   Plan

Check   Do

Such tracking systems need not be costly to execute and maintain, but must be appropriately designed and executed.

Process safety management programs have come a long way in identifying major weaknesses in existing operating systems. However, some challenges remain:

- Comprehensive solutions to strengthen process safety systems may require a more thorough understanding of the potential multiple factors and their interaction that contribute to process safety incidents than can be ascertained from experiential observation alone.

- It is often difficult to develop a comprehensive strategy and plan without data-based analysis to accurately pinpoint areas of need and estimate resource needs.

- Justifying financial support for improving process safety systems often requires priority over competing business systems needs, and experienced opinion alone is not always sufficient justification. Company leaders are usually more influenced by data-supported justifications.

- Focusing upon a single improvement area, the "squeaky wheel," may distract emphasis or divert resources from equally important system performance deficiencies that are less obvious.

- A company history that includes few serious process safety incidents can lead to complacency and/or misplaced overconfidence without a means to measure process safety performance reliably.

Collecting the correct information to monitor and understand performance of the process safety management system, and its separate elements, can provide an unbiased and comprehensive view of system performance. This will alert the appropriate personnel to weaknesses in the process safety system. A company-wide versus a site-specific approach to process safety auditing, evaluation, and improvement not only deals with better prioritization of identified improvement work but potentially provides better auditing and evaluation against those conditions that one site may accept as "normal/acceptable" over time.

2.2.1   Occupational Safety Improvement

Many examples within the process industries in which collecting metrics have spurred management commitment to improve performance. Though there are many differences between managing personnel safety and process safety, the collection of personnel safety data is a valuable example of how personnel safety improved once actual performance was understood at a high level. Many years ago, the U.S. OSHA regulated workplace record keeping and the reporting of normalized occupational injury and illness (OII) rates that allowed comparison of rates within and across different sizes and types of firms (29 C.F.R. §1904). Even though there was no government mandate to use such records to force improved performance, or lower OII rates, many companies tracked the OII rates and monitored the rates internally to understand if OII rates were decreasing or increasing over time. Some were surprised to learn that their rates were not among leaders in their industrial sector, and in some cases, their OII rates had increased. This data prompted companies to question why certain trends existed and how their safety records could be improved.

In parallel, trade organizations collected the OII data from members and provided member-to-member comparisons. Many companies that found themselves in an undesirable position (less-than-average performance) used the rate data to justify implementing improvement programs and expanded their own data collection efforts to gain a deeper understanding of their performance in personnel safety. The natural competitiveness in human nature also played a role in prompting companies to achieve significant reductions in their OII rates as compared to their peers. Some companies achieved a several-fold reduction in OII rates, and some multinational companies implemented similar data collection programs across their worldwide operations. The benefits of such programs went beyond a simple improvement of OII rates, materially lowering the number of employees injured each year and cost associated with those injuries (medical expenses, lost productivity, etc.). This supported an improved safety culture among the employees, inspiring the work force to fully engage in the performance improvement efforts.

Even though there appears to be no correlation between occupational safety and process safety performance (Elliott et al., 2008), process safety improvement efforts need to follow the same philosophical directions as the above example— collect the proper information and devise improvement plans. Capturing the commitment of employees and modifying culture can be difficult until employees actually see and understand the benefit from the efforts.

2.2.2   Reliability Improvement Programs

Reliability programs are a relevant example of how data-based "quality systems" have improved operating efficiency and cost effectiveness in the process industries. The reliability programs have also had a positive effect upon process safety since the reliability goal is to prevent unanticipated equipment failures, many of which can lead to a unit shutdown and/or loss of containment that could result in a process safety incident. Some companies have moved from a reliability effort primarily justified by maintenance cost savings to a system with the objective of maximizing the processing facility productivity by avoiding unscheduled shutdowns and achieving superior equipment reliability, often called increased "asset utilization."

The benefits to process safety are obvious since eliminating equipment failures removes potential causes for losing control of the process, losing containment of hazardous materials, and the necessity for nonroutine operations. Successfully implementing reliability programs depends upon collecting the proper data to assess performance and maintain reliable, dependable practices. Superior process safety performance also depends on maintaining reliable, dependable practices and can benefit from a similar approach.

2.2.3 Benchmarking

Process safety data facilitates performance comparison among process industry companies or among plants within the same company. Such data comparison, often called benchmarking, is promoted as a means to better identify opportunities for performance improvement. It is essential for companies or industries involved in benchmarking to use common metrics. Chapter 7 describes how an industry organization can demonstrate its members' good performance, and shows that good process safety performance is a very positive result. Benchmarking is also valuable among those who share their process safety performance results with their peers to learn new approaches to improving process safety and to understand if their approach and performance is markedly different. Fortunately, the practice of sharing process safety information transcends the normal market competitive forces and is viewed as a means to improve safety for everyone. Within the process safety industries, any significant process safety incident reflects poorly upon the entire industry, so sharing process safety information benefits everyone and reasonably leads to improved performance. Professional societies and industry groups provide forums to share process safety performance information among member companies.

2.3 TRACKING OPERATIONAL PERFORMANCE AND PROCESS SAFETY PERFORMANCE

All activities within the process facility may have an impact on process safety performance. Many operational activities are necessary to carry out the purpose of the processing unit (producing the desired product), and dependable execution of these activities is necessary to sustain acceptable process safety. There is a strong link between the operational discipline of a unit and the process safety performance of that unit. By tracking the appropriate indicators or metrics in a processing operation over time, the performance can be monitored and evaluated against targets. Metric

> *If you are not managing process safety well, you are probably not managing other things well.*

analysis can provide insights to judge the effectiveness of a process safety management system. Note that many of the indicators tracked by an operating unit are not solely defined for process safety management purposes, but relate to how well the processing operation is conducted relative to production and quality targets. Many indicators can serve the dual purpose of tracking operational performance as well as process safety performance.

Several process safety elements do not lend themselves to casual observation. A disciplined assessment over time is necessary to ascertain the true system performance. Tracking factors such as frequency of exceeding safe operating limits, not completing inspections on time, not conducting the proper management of change reviews, employees' not completing scheduled training on time, and so on provides information on the health of the process safety system. While the objective is to ensure adequate performance of the process safety systems, often the performance must be measured over an extended period to develop an accurate picture of performance. A one-time glitch within a process safety element does not indicate the system is in trouble, nor does the lack of major process safety incidents indicate the process safety system is robust.

Selection, and the number, of the indicators as metrics are important since some indicators may not provide the needed insights to ensure acceptable performance. Poorly selected metrics can produce gaps of knowledge and may result in unwarranted overconfidence. While tracking personnel injuries is appropriate and valuable for occupational safety programs, tracking such incidents has little relevance in assessing process safety performance. The Baker Report on U.S.–based BP Refineries points out that a sense of overconfidence in a good occupational safety record can contribute to losing sight of important process safety system deficiencies (Baker, 2007).

2.4 AVOIDING COMPLACENCY

Collecting and evaluating the metrics can help identify problems within the process safety management system as well as the process safety culture of the organization. Acceptance of chronically overdue inspections, not precisely following operating procedures, missed employee training, inadequate management of change reviews, and overdue process analysis studies may reinforce acceptance of deviation from procedures as the norm: "We didn't follow the operating procedure but nothing happened, so it's OK." This can lead to complacency and affect the drive toward reliably executing operating and process safety systems. Accepting deviation and failing to respond to near misses can be seen as normalizing unintended behaviors and conditions, and this can lead to a gradual degradation of performance that eventually may give rise to a dangerous situation.

2.5 CONCLUSION

Achieving high-quality process safety management is critical to any enterprise that operates hazardous processes. Employees, shareholders, and the general public expect facilities that store, use, manufacture, or ship hazardous materials will take the steps necessary to manage those operations safely—because the impact of failure is too high. Serious process safety incidents not only injure or kill employees and members of the local community and shut down production, but also can result in significant costs to the organization to resolve claims and fines as well as the potential for criminal prosecution. Such incidents not infrequently generate the promulgation of new regulatory requirements and may even threaten the very existence of a company.

Process safety metrics are a significant component of any process safety management system. Such systems are critical mechanisms for reducing process safety incidents by measuring process safety performance both to identify performance strengths and weaknesses and to achieve and maintain superior performance. Properly selected metrics that fit with the detailed objectives of an organization will identify the successes and point out the weaknesses of the system. The Plan-Do-Check-Act cycle is a useful model for continuously improving the performance of the process safety management system and the metrics that undergird the system.

REFERENCES

Baker, J.A. et al., *The Report of the BP U.S. Refineries Independent Safety Review Panel,* January 2007, *http://www.bp.com/liveassets/bp_internet/globalbp/ globalbp_uk_english/SP/STAGING/local_assets/assets/pdfs/Baker_panel_ report.pdf*

BP, "Annual Report and Accounts 2007," February 22, 2008

Buncefield Major Investigation Board, *The Buncefield Incident 11 December 2005:The Final Report of the Major Incident Investigation Board,* Crown, December 2008

Center for Chemical Process Safety, *Guidelines for Engineering Design for Process Safety,* American Institute of Chemical Engineers, New York, 1993

Center for Chemical Process Safety, *Guidelines for Process Safety in Batch Reaction Systems,* American Institute of Chemical Engineers, New York, 1999

Center for Chemical Process Safety, *Guidelines for Risk Based Process Safety,* American Institute of Chemical Engineers, New York, 2007

Center for Chemical Process Safety, *Incidents that Define Process Safety,* American Institute of Chemical Engineers, New York, 2008 (CCPS, 2008a)

Center for Chemical Process Safety, *Inherently Safer Chemical Processes: A Life Cycle Approach, 2nd Edition,* American Institute of Chemical Engineers, New York, 2008 (CCPS, 2008b)

Chemical Week, "Former Concept Sciences President Indicted on Criminal Charges," March 3, 1999

Elliott, M.R. et al., "Linking OII and RMP data: does everyday safety prevent catastrophic loss?" *International Journal of Risk Assessment and Management,* Vol. 10, Nos. 1/2, 2008

U.S. Chemical Safety & Hazard Investigation Board, *Investigation Report: Refinery Fire Incident, Tosco Avon Refinery, Martinez, California, February 23, 1999,* Report No. 99-014-I-CA, March 2001

3

PROCESS SAFETY MANAGEMENT METRICS

Process safety metrics are critical indicators for evaluating a process safety management system's performance. Tracking the number of process safety incidents is one common measure of performance, but merely tracking the number of incidents after the fact is insufficient to understand the system failure that allowed the incident to occur and what can be done to prevent a recurrence. More than one metric and more than one type of metric are needed to monitor performance of a process safety management system. A comprehensive process safety management system should contain a variety of different metrics that monitor different dimensions of the system and the performance of all critical elements.

> *A "process safety metric" is an observable measure that provides insights into a concept — process safety — that is difficult to measure directly.*

Different metrics may be used to describe past performance, predict future performance, and encourage behavioral change. They are a means to evaluate the overall system performance and to develop a path toward superior process safety performance. This is accomplished by identifying where the current performance falls within a spectrum of excellent-to-poor performance. Such information will allow executives and site management to develop plans to address the specific improvement opportunities that could lead to measurable improvement in process safety. Good process safety metrics reinforce a process safety culture that promotes the belief that process safety incidents are preventable, that improvement is continuous, and that policies and procedures are necessary and will be followed. Continuous improvement is necessary and any improvement program will be based on measurable elements. Therefore, to continuously improve performance, organizations must develop and implement effective process safety metrics.

3.1 METRICS AND THE PROCESS SAFETY MODELS

Serious process safety incidents involving loss of life, serious injury, or property damage are often preceded by multiple less-serious incidents or near misses as well as recurring unsafe behaviors or insufficient operating discipline. Process safety metrics are established, collected, and acted upon to help companies and industries improve their performance by identifying problems, failures, and lapses that can lead to process safety incidents or events as well as to track performance over time.

The process safety pyramid described in Figure 1.2 provides a useful concept for categorizing metrics by severity. *Lagging, leading, and near-miss* metrics are associated with the different levels of the safety pyramid. Figure 3.1 illustrates how each of these four areas fit into that categorization.[5]

Figure 3.1 Metrics and the Process Safety Pyramid

Lagging
Metrics

Near Miss Metrics

Other Near Miss

Leading Metrics

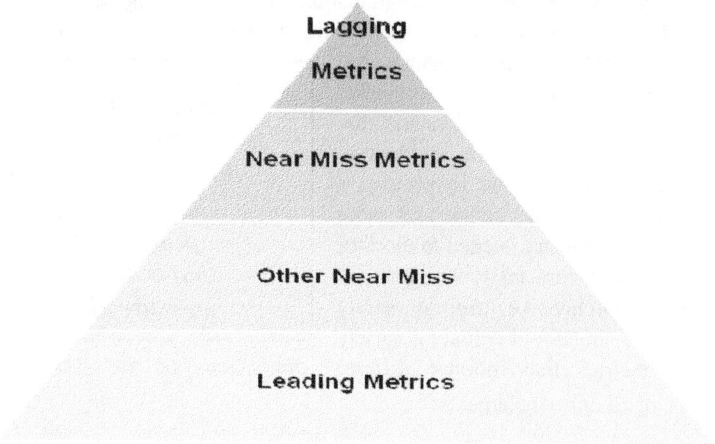

Lagging Metrics—process safety incidents that meet the threshold of severity and should be reported as part of the process safety metric

Near-Miss Incidents—incidents that did not meet the definition of process safety incident metric

Other Near Misses—system failures that could have led to an incident

Leading Metrics—measurements to ensure that safety protection layers and operating discipline are being maintained, including unsafe behaviors or insufficient operating discipline equipment selection, engineering design, specification of inspection frequency, and technique

Identifying and correcting behaviors or problems at the lower levels of the pyramid will help eliminate paths to an incident at the top. Failure to learn from near misses and to identify and remedy systemic flaws can produce catastrophic results.

The Swiss cheese model (see Figure 1.3 in Chapter 1 and Figure 3.2) offers another useful way to envision how failure to contain a hazard could result in a serious consequence if the process safety barriers and controls have weaknesses—the holes in the slices of Swiss cheese. For example, in order to prevent the risk of an accident, such as hydrocarbon release, a number of barriers (i.e., risk reduction

---

[5] The pyramid is divided into four layers to emphasize two different types of near misses.

measures in the management system involving plant, process, and people) are established. Further "escalation controls" are also in place to respond to the "top event," such as fire and gas detection, shutdown systems, and emergency response to control and/or mitigate the consequences of the top event. If consequences do result, then this demonstrates that all the barriers have holes (weaknesses).

Figure 3.2 Swiss Cheese Model (CCPS, 2007b)

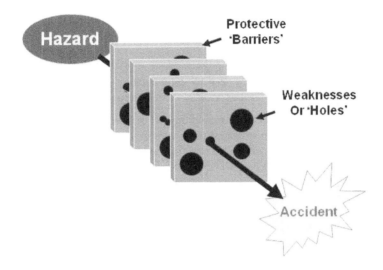

Regardless of whether one uses the process safety pyramid, the Swiss cheese model, or something else (for example, the anatomy of an incident model discussed in HEP 3), the concepts of lagging, leading, and near miss exist on a continuum without strict definition that places them in one category as opposed to another. A lagging metric for an outcome that is serious but not catastrophic may be a leading metric for an accident or event with more serious consequences. Further, some incidents viewed as lagging near misses, where the consequences are low, are often described as "leading" even though the data point is recorded after the near-miss incident.

> *Regardless of whether a particular metric is classified as lagging, leading, or a near miss, the purpose of the metric is to establish good indicators of conditions that could ultimately lead to a severe event.*

Companies seeking to develop a suite of lead and lag indicators may find it impossible to decide whether particular indicators are leading or lagging, and they may worry about whether they have an adequately balanced suite of indicators.

But they should not worry: The important point is not to develop a crisp definition to delineate the differences among the terms, but to capture and document that a failure of some sort occurred (Hopkins, 2009). Regardless of whether a particular metric is classified as a lagging indicator, a near miss, or a leading indicator, the purpose of the metric is to establish good indicators of conditions that could ultimately lead to a severe event.

3.1.1 Lagging Metrics

Lagging metrics are outcome-oriented and retrospective; they describe events that have already occurred and may indicate potential recurring problems and include fires, releases, and explosions. Lagging metrics should be easily and objectively measurable. They may be differentiated according to severity and, when broadly accepted and reported, used in company- or industry-wide benchmarking and comparison. Lagging metrics are used by companies, industries, and the public as the ultimate measure of whether a process safety management system is working.

Criteria for lagging metrics should be set below the level that describes catastrophic events. While it is important to capture those catastrophic incidents, it is equally, if not even more, valuable to define such metrics to capture less-severe incidents, including the failure of the process safety elements, which protect against or limit the consequences of a major incident (HSE, 2006). Lagging metrics that represent failures of a less-severe nature are valuable in identifying potential hazards before a catastrophic event occurs. Such information helps management understand issues at an early stage when they may be more easily addressed.

Because they are a retrospective measure, learning from such metrics can come only after an accident or failure. A simple accounting of catastrophic events alone does not provide sufficient information since the incidents are infrequent and with disparate causes. While it is important to track such incidents, it is also important to track other, less-severe incidents and near misses.

3.1.2 Near-Miss Metrics

Near-miss metrics include actual process safety incidents that do not meet the threshold defining a lagging metric as well as system failures that could have, but did not, lead to an incident such as an instrumentation failure or vessel corrosion. Near misses are often less obvious than accidents and are defined as having little if any immediate impact on individuals, processes, or the environment. Despite their limited impact, near misses provide insight into accidents that could happen.

Phimister recommends a broad near-miss definition that can be easily understood by all employees, one that focuses on identifying a situation from which site environmental, health, and safety (EHS) performance may be improved:

An opportunity to improve environmental, health and safety practice based on a condition, or an incident with potential for more serious consequences (Phimister, 2003).

Commonly reported near misses include such events as exceeding operating limits, a release of a chemical or other hazardous substance that does not meet the threshold for a process safety incident metric, activation of relief valves, interlocks, or ruptured disks. Companies may establish near-miss metrics based on the specifics of their operation, based on their observations of frequent upsets or failures, or to track and correct observed unsafe practices or behaviors.

A near miss may be considered either a lagging or leading indicator, or both. While it does describe an event that has already happened (lagging), it can also be considered an indication of a more hazardous situation (leading). Regardless of how they are labeled or organized, identifying and investigating near misses is important. Near misses can provide valuable information that can be used to address a risk before a more serious incident occurs. While it is important to link failures or weaknesses identified by near misses, as well as incidents, to the specific failure, it is also important to examine both the actual result of a near miss and the most credible potential result.

Near-miss incident reporting is often an internal metric since the triggering events may occur more frequently but with lower consequences. An increase in near-miss reports should be expected at first, but consistent reporting should lead to a decrease in more serious process safety incidents if findings and recommendations of near-miss reports are acted upon. In fact, management should not express or infer the view that near-miss report rates are a desirable metric to decrease over time (Phimister, 2003). Employees will become more comfortable reporting near misses when they understand that the consequence of such reporting can constitute a significant savings compared with the potential damage, down time, and loss of reputation from a serious incident. Developing and choosing near-miss metrics is discussed more fully in Chapter 4.

3.1.3   Leading Metrics

Leading metrics are forward-looking and indicate the performance of the key work processes, operating discipline, or layers of protection that prevent incidents. They are designed to give an early indication of problems or deterioration in key safety systems early enough that corrective actions may be taken to avoid a major process safety incident.

The percentage of equipment overdue for inspection can be considered a leading metric as it can relate to the physical condition of the equipment as well as the robustness of the inspection program (and, by extension, indicate a gap in the safety culture). Of course, even if inspections are routinely conducted and recorded as a leading metric, a failure can still occur. This reinforces the need for multiple process safety metrics to address positive signs of systems working well or negative signs of failures of various layers of protection. Mechanical integrity (MI), action item follow-up, management of change (MOC), process safety training, and process safety culture are examples of process safety elements that could be a source of leading metrics.

3.2   OTHER METRIC DIMENSIONS

In addition to the continuum of lagging–near-miss–leading, metrics may be described by additional dimensions based upon *what* is actually measured (an activity or quality of performance) and *who* will use the information collected by the metric (internal or external audience or both).

3.2.1   Activity and Outcome Metrics

In addition to differentiating among the severity of a metric (incident, near miss, unsafe behavior), a metric can track *activities* (whether an action occurs) as well as the *outcome* (the quality or performance) of the action.

- *Activity metrics* are pro-active leading metrics that measure how well a facility is meeting the requirements of an established system. Activity metrics can track, for example, the number of scheduled risk assessments completed, the percentage of scheduled operating procedures validated, and the number of scheduled emergency drills completed. A start-up system can track the percentage of completed activities building towards a new system or a revision, e.g., the percentage of piping and instrumentation diagram (P&ID) walked down to verify their accuracy and the percentage of sited buildings as required by building siting guidelines. Activity indicators provide organizations with a means of checking, on a regular and systematic basis, whether they are implementing their priority actions in the way they were intended and why a result (e.g., measured by an outcome indicator) has been achieved or not (OECD, 2008).

- *Outcome metrics* assess whether safety-related actions (policies, procedures, and practices) are achieving their desired results and whether such actions are leading to a decreased likelihood of an accident occurring and/or lower consequences from an accident. Sometimes referred to as performance, effectiveness, or quality metrics, they are reactive, lagging metrics that measure changes in safety performance over time, or failure of performance. For example, a change in the number of leaks or process fires could indicate the activities (MOC, inspections, etc.) were being done and were being done well. Outcome indicators show *whether* a desired result or adherence to requirements has been achieved but, unlike activities indicators, they do not explain *why* the result was or was not achieved (OECD, 2008).

Both activity and outcome metrics are needed to understand fully the performance of the system—that a particular activity is being conducted and the outcome or quality of the activity. For example, the system may measure the activity of carrying out MOC—how many are conducted and their frequency. The audit of the MOC system coupled with a change in the number of incidents or near misses associated with the MOC system describes the outcome or quality of the MOC program.

3.2.2   Internal and External Metrics

A final dimension of a metric is the intended audience. Information may be disseminated to *internal* audiences such as employees, supervisors, facility management, senior executives, and the board of directors or to *external* audiences such as a trade association, regulatory agency, or the general public. Any metric that must be reported for regulatory compliance is, by definition, an external metric. Chapter 6 discusses in greater detail the identification of metrics for various external and internal audiences.

- *Internal metrics* are those used within an organization to manage an effort or activity. Such metrics should be readily available within, but not necessarily distributed outside, the organization. Internal metrics will include leading, lagging, and near-miss metrics as well as metrics that track activity and outcome. Internal metrics can include tracking whether employees have received and understood training, whether policies are updated and conveyed to employees and contractors, and the extent to which procedures are followed and employees' suggestions are considered. Near-miss metrics are generally considered internal metrics since they are rarely shared outside the organization. Internal metrics may be used to develop external metrics as well as other information for external audiences.

- *External metrics* are those reported broadly within and outside the organization to demonstrate publicly the organization's performance. For example, reports to government agencies (e.g., injury and illness rates, environmental emissions) are external metrics. Also, members of the American Chemistry Council (ACC) annually report process safety performance metrics as part of mandatory obligations under the Responsible Care® program.[6] In Europe, the Major Accident Reporting System (MARS) provides open access to information on major accidents as part of compliance with the Seveso II Directive.

---

[6] Responsible Care® is the chemical industry's global voluntary initiative under which companies, through their national associations, work together to continuously improve their health, safety, and environmental performance, and to communicate with stakeholders about their products and processes.

Both internal and external metrics are valuable. Internal metrics provide information to those throughout the organization with the information needed to evaluate the progress and effectiveness of the process safety management system. External metrics allow outside stakeholders to evaluate the organization's performance and to hold those within the organization accountable for unacceptable performance. (See Chapter 7 for a discussion on accountability.)

3.3 FORMS OF METRICS

There are many forms a metric can take. Some can be expressed in *absolute form*, while others are expressed in *ratios* or *indices* that provide context and enable more effective comparison over time and across the organization. The form of a metric contributes to how meaningful and useful the metric is for its intended uses and users. In determining the form and what is included within the metric, the organization also needs to determine whether the metric can be meaningful and useful at the different organizational levels.

3.3.1 Absolute Metrics

An absolute measure is one in which a simple count of events is recorded and reported. This would include the number of occurrences of a specific type of incident per year (e.g., release, loss of containment, fire) or activity (e.g., how many employees were trained, whether a mechanical integrity evaluation was conducted). Absolute metrics do not necessarily provide information on the quality of an activity or a change or trend in the activity over time. Absolute metrics also may be difficult to compare meaningfully across the organization. However, regulatory agencies and the public may be very interested in some absolute metrics. For example, the public wants to know the number of releases from a facility and the quantity of the release—particularly from a facility in their community. However, how a given release or a facility's release record compares with facilities outside the community may be of minimal, if any, interest.

3.3.2 Normalized Metrics

Normalized metrics can produce ratios that provide better context for comparison across multiple sites, companies, or industry segments. Such metrics can also be useful in one-to-one performance benchmarking. Normalized data provides a "rate" expression for process safety performance and can be used to compare performance among sites that differ in size and processing technology as well as performance comparisons among companies within an industry or even across industries. There are a number of ways to normalize process safety data, and methods include, but are not limited to:

- The number of personnel hours worked during a defined period
- Total production volume produced during a defined period

- The number of lots produced during a defined period
- The percentage of mechanical integrity inspections performed on time
- Other normalizing factors that the organization deems appropriate.

Normalizing data may also use factors associated with process hazards or risks, manufacturing process complexity, and other factors that a company may feel are relevant to understanding their process safety performance.

Various industry organizations have debated the appropriate normalizing factor for process safety data. Even though many normalizing factors have merit, no clear consensus has been reached. In "Process Safety Leading and Lagging Metrics" (CCPS, 2007b), the Center for Chemical Process Safety recommends using the U.S. OSHA–defined normalizing factor of 200,000 hours worked per year. This factor represents the total hours worked by 100 workers for a full year. This normalizing factor is well understood and has been adopted by other countries for occupational injuries. It is appropriate to use 200,000 hours worked by personnel involved in the processing operations for normalizing process safety data since this represents those persons who could be potentially impacted by an incident.

Using *only* normalized data to evaluate process safety performance should be approached cautiously. Normalized data presentations, by focusing on rates, may draw focus away from the details and lessons from individual incidents. Normalized data, such as the number of near misses per quarter or incidents per year, can be extremely valuable in placing incident information into a context or allowing comparisons among data sources or over time, but such information cannot substitute for lessons learned by investigating all incidents.

> *Using only normalized data to evaluate process safety, performance should be approached cautiously.*

Normalized performance data can facilitate sharing performance results both internally and externally. Using a consistent metrics basis, normalized data can aid in performance comparisons across an organization that consists of multiple sites and facilities. At the site level, multiple units can be compared using metrics of interest. Several industry and professional organizations sponsor the sharing of normalized metrics to determine the overall performance trend by industry segments.

3.4 CHARACTERISTICS OF GOOD METRICS

Some attributes of a process, such as temperature and pressure, are directly measurable. Others, such as attitudes of employees or the public, may only be measurable indirectly. Much thought and consideration must be given to developing those metrics that will need to be compared across the organization and will drive correct behaviors without masking others. For example, overemphasis on OII can drive occupational safety reporting and improvement, but will do little to improve process safety.[7] A good metric should:

- Allow accurate and detailed comparisons
- Lead to correct conclusions and avoid erroneous conclusions
- Be well understood
- Have a quantitative basis.

To be useful and credible, a metric must exhibit certain characteristics or meet certain criteria.

- Metrics must be reliable. They should be measurable using an objective or unbiased scale. This means that the person designing the metric and using the information is confident that the metric is a true measure of what is happening. To be measurable, a metric must be specific and discrete.

- Metrics must be repeatable. Similar conditions will produce similar results, and different trained personnel measuring the same event or data point will obtain the same result.

- Metrics must be consistent. The units and definitions are consistent across the organization. This is particularly important where metrics from one area of the organization will be compared with those of another.

- Metrics must be independent of outside influences. Much will be riding on the performance described by the metrics—from personnel evaluations of those responsible for certain operations to the lives of employees and the surrounding community. The metric must lead to correct conclusions and be independent of the pressures to achieve a specific outcome.

- Metrics should be relevant to the behavior or process being measured; they must have a purpose and lead to actionable responses when the results of such behavior or process are found to be beyond the target or control limits. As discussed in later chapters, a metric system will evolve as data is collected and evaluated. Data collected over time is likely to change as practitioners become more knowledgeable and improvements in systems are acknowledged, thus developing and enhancing the relevance of the metric.

- Metrics should be comparable with other similar metrics. Comparability may be over time, across a company, or across an industry on a global basis. Consensus metrics within a company allow for comparison across facilities and operations. Consensus metrics at the industry level—such as

---

[7] In 2008 Michael Elliott and colleagues at the University of Pennsylvania Wharton Risk Center published a study ("Linking OII and RMP Data: Does Everyday Safety Prevent Catastrophic Loss?") that provided no evidence of a correlation between low OII rates and low process safety incident rates and only marginal support for a correlation that high OII rates might predict further high-consequence process safety events.

those developed by trade associations, professional organization, or intergovernmental groups—allow comparisons within and across companies or industries.

- Metrics should be appropriate for company or regulatory compliance. Some government regulations prescribe specific tests and parameters— such as occupational injuries or illnesses or environmental releases—to be tracked and reported to regulatory agencies and the public. Organizations may adopt regulatory metrics as part of their internal systems to avoid redundant reporting, ease regulatory reporting, and allow benchmarking with similarly regulated companies. Organizations may also identify specific metrics that will be used to evaluate system and individual performance against requirements.

- Metrics should include sufficient data to be meaningful, i.e., statistically sufficient in order to measure positive or negative change. The most catastrophic process safety incidents are rare occurrences, but contributing factors are often present in many types of incidents, including near misses. Database volume is increased by collecting data on elements and sub-elements of the process safety system. If metrics are based on a rate, sufficient instances of the events must be counted to be able to speak meaningfully about a rate. Weakness in elements of process safety (MOC, process knowledge, training, etc.), tracked over time and collected over an enterprise, can indicate endemic weakness in a system.

- Metric should be appropriate to the audience. The data and metrics reported will vary depending upon the needs of a given audience. While information for upper management usually contains aggregated or normalized data, focuses on trends, and is only provided on a periodic basis (such as quarterly or annually), data for operating personnel is generally more detailed and reported more frequently.

- Metrics should be timely, providing information when needed based on the purpose of the metric and the audience receiving the metric. Shorter-term metrics include those designed to identify conditions that need immediate attention such as the number of instances when the temperature control is outside safe operating limits and how far outside the limit. Longer-term metrics are those that measure more gradual changes in performance, changes that can only be assessed over a lengthier period of time such as performance trends. A safety culture survey tracking employee attitudes and shared beliefs, norms, and practices is a good example of a long-term metric, one where the data gathering may be over long time spans, e.g., annual or biannual. Short-term metrics may be rolled into longer-term metrics. For example, a facility may track leaks on a variety of schedules—weekly, quarterly, and annually. Investigating individual leaks will identify root and contributing causes, but only by examining leaks over time can an one estimate whether the leaks are performance anomalies or a trend indicating a more systemic problem.

- Where possible, metrics should be easy to use. Metrics that are hard to measure or derive may be less likely to be measured or less likely to be measured correctly. A wrong value is worse than a bad metric! However, not everything that needs to be measured or monitored will necessarily be easy to use, which can be an intrinsic conflict between accuracy and understandability of metrics. For example, the occupational injury and illness rate has been so effective because it is simple and easy to understand, even for nonprofessionals. However, it does not take into account complexities such as the big variation of hazards and activities of different workplaces.

- Metrics should be periodically audited to ensure they meet all the above expectations. For example, the near-miss data should be reviewed to ensure the appropriate investigations were done. Operating logs should be reviewed to ensure all near misses were reported, investigated, and recorded.

3.5   CONCLUSION

Process safety metrics are critical indicators for evaluating a process safety management system's performance. More than one metric and more than one type of metric are needed to monitor performance of a process safety management system. A comprehensive process safety management system should contain a variety of metrics that monitor different dimensions of the system and the performance of all critical elements.

Lagging, near–miss, and *leading* metrics are associated with different severity levels of the safety pyramid (see Figure 3.1), but the difference among lagging, near-miss and leading indicators may not be distinct. That a failure is documented and evaluated is more important than how a metric is labeled. In all cases, the purpose of the metric is to establish good indicators of conditions that could ultimately lead to a severe event. In addition to severity, other metrics are valuable for tracking *activities* (whether an action occurs) and *outcome* (the quality or performance). Both activity and outcome metrics are needed to understand fully the performance of a system.

Thought and consideration must be given to developing those metrics that will need to be compared across the organization that will drive correct behaviors without masking others. A good metric should allow accurate and detailed comparisons, lead to correct conclusions and avoid erroneous conclusions, be well understood, and have a quantitative basis.

Good process safety metrics will reinforce a process safety culture promoting a belief that process safety incidents are preventable, that improvement is continuous, and that policies and procedures are necessary and will be followed. Continuous improvement is necessary and any improvement program must be based on measurable elements. Therefore, to continuously improve performance, organizations must develop, implement, and review effective process safety metrics.

REFERENCES

Center for Chemical Process Safety, *Guidelines for Hazard Evaluation Procedures, 3rd Edition* (HEP 3), American Institute of Chemical Engineers, New York, 2008

Center for Chemical Process Safety, *Guidelines for Revalidating Process Hazard Analysis,* American Institute of Chemical Engineers, New York, 2001

Center for Chemical Process Safety, *Guidelines for Risk Based Process Safety,* American Institute of Chemical Engineers, New York, 2007

Center for Chemical Process Safety, *Process Safety Leading and Lagging Metrics,* American Institute of Chemical Engineers, New York, 2007

Elliott, M.R. et al., "Linking OII and RMP Data: Does Everyday Safety Prevent Catastrophic Loss?" *International Journal of Risk Assessment and Management,* Vol. 10, Nos. 1/2, 2008

Health and Safety Executive (HSE), *HSG254 Developing Process Safety Performance Indicators, a Step-by-Step Guide,* Sudbury, Suffolk, UK, 2006

Hopkins, A., "Thinking About Process Safety Indicators," *Safety Science,* Vol. 47, No. 4, 2009

Organization for Economic Coordination and Development (OECD), *Guidance on Safety Performance Indicators related to Chemical Accident Prevention, Preparedness and Response for Industry, 2nd Edition,* OECD Environment, Health and Safety Publications, Series on Chemical Accidents No. 19, Paris, 2008

Phimister, J. et al., "Near-Miss Incident Management in the Chemical Process Industry," *Risk Analysis,* Vol. 23, No. 3, 2003

4

CHOOSING APPROPRIATE METRICS

Any organization that wishes to improve and sustain process safety performance must clearly establish its performance goals and objectives. Efforts to sustain and improve process safety performance will require a means to measure and evaluate that performance. Collecting performance information enables an organization to achieve their desired goals and objectives by helping to:

- Identify system improvement opportunities;
- Sustain performance at the desired level; and
- Measure progress toward performance improvement goals.

Without measuring performance, it will be very difficult, if not impossible, to gauge the success of improvement efforts or to detect gradual, complex trends within process safety system performance. Process safety metrics provide the data needed to evaluate and monitor performance.

This chapter goes through the process of developing and clarifying the process safety improvement goals for an organization. The first sections discuss formulating goals and objectives for the metrics system. Later sections discuss the factors to consider when selecting various metrics.

4.1 PROCESS SAFETY GOALS AND OBJECTIVES

As an organization begins an effort to improve process safety performance, it should establish specific goals and objectives that reflect performance outcomes and the desired future state of process safety. Senior leadership needs to lead the development of goals and objectives to signal full support and demonstrate active engagement.

Developing an effective performance improvement effort starts with defining goals and objectives.

Goals are usually high-level outcomes that describe a desired future state. Normally, the goals represent improved performance in an organization's activities and operations. For process safety, the ultimate goal is the elimination of all process safety incidents. Some organizations state this long-term inspirational goal as "zero process safety incidents." This overriding goal may not be achieved over a finite period of time, but the intent remains the same, providing a consistent directional vision for continually improving process safety. Some organizations describe such long-term goals as "values" that are considered cultural foundational elements, permeating every action taken.

Process safety improvement efforts will include performance goals that define the desired future state for the various elements of the process safety system. Examples may include 100-percent reporting of process safety incidents and near misses, 100-percent on-time completion of process safety training, and timely resolution for all hazards analysis recommendations.

Objectives normally relate to the strategies and tactics employed to make progress toward defined performance goals. Objectives are more detailed in nature than goals and describe desired, incremental progress over defined time spans, for example, a specific increase in reliability for a specific process safety element over a year. Since it is widely held that process safety incidents are caused by a convergence of multiple contributing factors (see discussion of the Swiss cheese model in Chapters 1 and 3), objectives usually address reliable execution of tasks within individual process safety system components. Metrics that measure performance to objectives may include percentages of hazards analyses performed on schedule, resolution of hazard analyses recommendations in a timely fashion, **or** percentages of personnel that did not complete process safety training.

> *"ZERO process safety incidents" is a powerful, long-term inspirational goal for the processing industries.*

Often the distinction between goals and objectives is a fine line that is not always clear, and goals for a performance improvement program may sound very similar to the objectives. Performance improvement efforts will have goals that define a future desired state of performance to provide the context within which the objectives will be defined. For example, a goal may be to achieve 100 percent of mechanical integrity inspections completed on time, while an objective supporting that goal may be to verify the actual timing of all mechanical integrity inspections and to record whether the schedule was met for each inspection. It may be helpful to remember this hierarchy:

> *GOALS*
>
> *feed*
>
> *OBJECTIVES*
>
> *feed*
>
> *STRATEGIES*

- A goal defines the desired future state.

- Objectives define desired incremental progress toward the goal.

- Strategies are defined actions expected to achieve the objectives.

This chapter will tend to use goals to describe a desired future state of performance. Objectives relate to strategies defined over the short term to improve performance. It is not important to make an absolute distinction between goals and objectives as long as performance measures are clearly defined and understood. However, the deliberation process to define goals and then define objectives often helps to identify the most appropriate strategies for the improvement effort.

4.2. DEFINE THE PROCESS SAFETY GOALS

Organizations typically define high-level process safety system goals that reflect their principles, policies, and values. Appropriate metrics are selected to monitor progress toward reaching these goals (the objectives). For example, the overall goal for eliminating all process safety incidents is likely to use as the performance metric the number of process safety incidents that occur annually. The purpose of the process safety metrics system is to monitor and evaluate performance and to ensure performance is moving toward meeting the defined goals. Metrics also provide the information required to know whether desired performance is being maintained and not declining.

Process safety goals apply to all levels of the organization. Goal setting usually starts at top levels of an organization and flows down through the ranks. Typically upper management provides simply stated goals, such as "eliminate process safety incidents," and these high-level goals provide the context for goal setting in lower levels of the organization. Such visionary goals also need to be defined in a way that facilitates translating goals into actionable objectives. Personnel involved directly with the operation of the process safety systems will define goals to improve system performance in ways that contribute to meeting the higher-level goals. Those responsible for the process safety management system will establish goals to improve the reliability of the process safety system elements, following the premise that reliable system elements will decrease the number of process safety incidents. One system goal could be to investigate 100 percent of the process safety incidents. This goal may utilize multiple metrics to evaluate performance of the incident investigation element, such as the number of recorded incidents, the number of investigations completed, and the number of unresolved investigation recommendations. Goals for all levels in the organization need to be consistent in purpose to achieve improved performance across an organization.

4.2.1   Periodically Review Process Safety Goals

Companies committed to process safety improvement review their process safety goals on a regular basis. For instance, progress toward achieving process safety goals can be reviewed during the annual corporate-planning exercise along with other business goals. Integrating process safety goals with corporate and business goals keeps the process safety goals fresh and visible within the organization. This helps drive improvement efforts through regular reinforcement and commitment to improved process safety. Business units often develop specific process safety objectives to address recognized issues unique to their operations. For example, a business that processes and handles combustible dusts may strive to drive down the number of process safety incidents associated with combustible dusts.

4.2.2   Relate Goals to Process Safety Elements

For the process safety system to meet performance expectations, system components need to function as intended. A small number of episodic system failures may seem tolerable, but **generally** as more systematic failures occur, the likelihood of suffering a process safety incident will increase. For instance, most process safety systems and process engineering designs contain multiple layers of protection, so individual system failures are not likely to lead to a catastrophic incident. However, if multiple safeguards do not exist or the safeguards are compromised over time, a finite risk can exist for a single failure to result in a process safety incident. Over the long term, routine failures in system components may provide opportunity for diverse contributing factors to culminate into a process safety incident. Process safety goals often address a reduction of failures within individual system elements. It is important to address and correct recurring component failures because as failures become routine, complacency toward failures and a pervasive attitude that accepts or normalizes deviations can develop.

The sum of failures affecting a process safety element can amount to a surprisingly large number of individual occurrences when all failures are tracked. For example, if scheduled relief valve inspections are routinely missed in a large facility, the total number of system failures could add up to hundreds of overdue inspections. If the number of system failures is large, it is unlikely that the total number of failures can be eliminated over a short time. It will be important to define short-term goals that address the most important areas first. Using the relief valve inspection example, the short-term goals should recognize that reliable performance of some relief valves is more critical than others. Achieving long-term goals for inspecting relief valves will follow a phased approach that sets priorities for addressing the most important system failures first.

Goal development for the various process safety system elements should consider the importance of individual components within the elements as they relate to overall system performance. System goals should focus upon system components that need to be strengthened and work to provide high reliability for all safeguards. Not all systems will have the same weaknesses and strengths, so goals are fashioned to address the needs of the given situation.

4.2.3   Periodically Review Goals and Priorities

An organization will follow various strategies to strengthen its process safety system. While developing the plans to improve process safety, the organization may identify many improvement opportunities. It is often impractical to address all identified opportunities simultaneously since the sheer number of potential failures would quickly overwhelm an organization's resources and capabilities. Often a phased approach is used in setting goals that address improving the highest-priority areas first. As performance improvement progress is realized, the goals should be revised to acknowledge and monitor performance gains, and to emphasize the next highest priority areas. Some find it helpful to maintain an official summary of improvement ideas that personnel suggest during system operations; these ideas are evaluated periodically for adoption. An organization will periodically redefine and refine its ongoing goals and priorities during implementation of the process safety performance improvement plans.

4.3   DEFINE PROCESS SAFETY OBJECTIVES

After an organization establishes process safety system goals, discrete objectives are defined and performance improvement strategies are developed. The organization should carefully define its detailed objectives that will have the greatest impact upon performance. These objectives need to be reviewed periodically and revised as performance changes over time. Often, the plan to improve process safety performance is implemented as a multi-phased set of objectives that address the highest areas of concern first, and once improvement is realized the focus shifts to the next highest priorities. Objectives may be identified using inputs from a number of sources including the experience of process safety professionals and operating personnel, analysis of process safety risks, regulatory requirements, and long-term improvement goals.

4.3.1   Define Objectives Based Upon Experience

Many organizations will decide that some areas of their process safety systems are not functioning as intended. This is often determined by recognizing that certain system elements are not executed reliably or that they are repeatedly identified as contributing factors in incidents. Indicators can be the number of process safety incidents or near misses or defined process safety system activities that are not carried out as intended. This evaluation is often based upon the collective judgment of the process safety professionals and operating personnel. When data exists to support this collective judgment, the evaluation is usually credible. In any case, the organization makes a decision to focus upon a specific system element(s), and experiential knowledge within the organization is a key factor that should guide where to focus improvement efforts.

4.3.2   Use Process Safety Risks to Define Objectives

The ultimate goal of the process safety system is to prevent process safety incidents. The Center for Chemical Process Safety's "Process Safety Leading and Lagging Metrics Report" (CCPS, 2007b) defines a consensus from several chemical and allied processing industries for definitions of process safety incidents and process safety near misses. If an organization adopts these definitions, a continuum of reported events, from minor events to catastrophic incidents, will exist. A minor incident may not result in injury, significant damage, or harm, but this information may indicate a potential exists for a more severe incident. Reviewing hazards and risk analyses can link specific failures in an operating process or safety system to potential consequences. This knowledge can provide a basis to set the objectives for the process safety systems performance:

- Stress highly reliable performance for areas with the potential to create a catastrophic event.

- Assign first priority to improving performance for high-risk areas.

As an example, higher priority and more time and resources should be applied to a phosgene manufacturing process than a water filtration system.

This is the basic philosophy detailed in the CCPS *Guidelines for Risk Based Process Safety* (CCPS, 2007a). Focus more priority on improving the elements that apply to the process areas with the highest potential process safety risk. Logically, more metrics (in addition to a base set of key metrics applied to all sites) will likely be used to track performance for those operations with the highest potential process safety risk.

4.3.3   Define Objectives Based Upon Regulatory Requirements

Some organizations may have to meet significant regulatory requirements. In cases where regulatory requirements are consistent with the organization's process safety objectives, successfully meeting the regulations may form the basis for establishing their objectives. Many facilities in the United States are subject to the U.S. OSHA Process Safety Management rule (29 C.F.R. §1910.119 (PSM rule)), and have used those regulatory elements to design their process safety management system. Tracking performance in meeting the PSM rule can provide an indication of the process safety system's strength and reliability. However, even if a facility is not covered by a regulation like the PSM rule, that facility may still have process hazards that warrant a process safety system. Some states and other jurisdictions within the United States have process safety requirements that may also be a source for metrics. Examples include:

- New Jersey Toxics Catastrophe Prevention Act (TCPA)
- Nevada Chemical Accident Prevention Program
- Delaware Risk Management Program
- California Accidental Release Prevention Program
- Contra Costa County (California) Industrial Safety Ordinance

Several countries have regulations detailing requirements for process safety. Countries in Europe have adopted regulations in response to the European Union's Seveso Directives. The Control of Major Accident Hazards (COMAH) regulation in the United Kingdom is one example. Brazil and Hong Kong have also promulgated regulations that relate to process safety risks. Some of these regulations do not define detailed programmatic requirements, but an organization can use these regulations as a starting point for developing a robust process safety management system.

4.3.4   Define Objectives Based Upon Long-Term Improvement Goals

An organization may focus on long-term goals as the basis for defining its short-term objectives. An organization may decide that the long-term view is a better perspective from which to design its performance improvement system. It is reasonable to define short-term, achievable objectives to demonstrate progress toward a worthy long-term goal such as eliminating loss-of-containment events arising from specific systematic failures. Achieving intermediate objectives demonstrates success of improvement strategies and allows tracking progress toward the long-term goal. Many organizations establish metrics to track progress toward the long-term goals, and these metrics are indicators of the overall performance for the process safety system.

4.3.4.1   Define Objectives Using System Strengths and Weaknesses

One of the first activities necessary to sustain and improve performance is to evaluate the strengths and weaknesses of the current process safety system. Every system will have some elements that are executed well and other elements where execution needs improvement. Determining the strengths and weaknesses can be accomplished using experience-based judgment or data that relates to system performance. Using data analysis is a good approach for evaluating system performance when appropriate data is available. However, some organizations may not yet have metrics in place or sufficient data for making data-based evaluations.

Where there are no metrics in place or insufficient data, weakness can be identified by using existing information, such as past audit results, incident causes, safety system failures, and recorded operation outside of safe operating limits. Most process upsets or product quality issues are indicators of process system weakness. The knowledge of operating personnel and process safety professionals can also provide an evaluation of the current system's strengths and weaknesses that can be used to formulate objectives. In the absence of a formal process safety system, surveying knowledgeable people within the organization who are knowledgeable regarding how current practices compare to the CCPS process safety elements can identify gaps that can be prioritized for action. Reliance upon experienced-based judgment of the process safety professionals and operating personnel provides a good start at identifying needed system improvements and defining objectives.

4.3.4.2   Determine What Works Well and What Does Not

Determine what has worked well in the process safety system. Answers to simple questions can provide a good indication of which process safety system elements work reliably and which elements are not working so well. A sampling of simple questions is listed below and is by no means a comprehensive list:

- Do budget and management commitments exist for reliably maintaining and operating the process safety system?
- Is a MOC review conducted for all process changes prior to implementing any change?
- Is all process safety information up-to-date?
- Are process hazards analyses (PHA) completed on time?
- Are all recommendations from PHA studies, incident investigations, and audits resolved in a timely fashion?
- Is operator refresher training completed on time?
- Are tests for safety critical process instrumentation completed on time?
- Are mechanical integrity inspections completed on time?

Developing detailed answers to these questions may not be easy, but once answers are developed, strengths and weaknesses in the process safety system will be identified.

4.3.4.3 Use a Baseline Survey

Some organizations begin the metrics selection process by conducting a baseline survey regarding the health of the process safety system. Such surveys may ask facilities to rate the performance of their process safety system performance using a self-audit format based upon the implemented elements. A simpler approach might be to ask operating managers what elements in their process safety system worry them the most.

4.3.4.4 Use Incident Investigation Results

The results of process safety incident investigations can be a rich source of information relating to the health of the process safety system. A properly conducted incident investigation will identify contributing factors, and evaluating these contributing factors can highlight weaknesses in the execution or design of the process safety system (or the process operation). Incident investigations can also determine if the performance indicated by collected metrics is accurate. For example, a failure or malfunction of a safety-instrument function (SIF) may indicate that scheduled inspection and testing for that SIF had not been carried out as intended. Metrics can be selected to monitor the SIF inspection and testing plan to ensure it is properly carried out. If incident investigations identify the same or similar contributing factors in multiple incidents, special attention is warranted for those factors. Process safety incident investigations may identify multiple contributing factors and provide the opportunity to focus improvement activities on several different process safety system elements. If 10 percent of reported process-related injuries are due to not using (or improperly using) personal protective equipment (PPE), then metrics that reinforce PPE use are appropriate.

> *A process that frequently challenges quality limits will eventually challenge the process safety limits.*

4.3.4.5 Use Process Safety Audit Results

Process safety audits are completed to understand how well the process safety management system is working. Process safety audits focus on the management system and execution of the process safety system tasks. Reviewing audit results can identify frequently recurring deficiencies and elements that require improvement, and evaluating the resolution of audit findings after closure (say, 6 to 12 months) can indicate if improvements are achieved. A potential drawback of the process safety audits is that the focus may only be on completing the management system tasks. Technical assessment of component design may not occur, and poorly designed elements will not accomplish the intended results. When using audit data, the metrics designer should determine if audit-identified performance deficiencies are due to failures of execution or to a component design flaw since understanding this difference will influence how objectives are structured.

4.3.4.6 Use Multiple Views for Validating Objectives

An organization may use various perspectives to validate their objectives. The factors for selecting objectives described above are related, but they differ in the thought process used for fashioning objectives. Often, organizations use a combination of approaches to define and validate their objectives. Using a different lens to review proposed objectives can provide valuable insights that either validate the objective as defined or highlight a worthy revision. Operating and craft personnel, supervisors, unit managers, site managers, and process safety professionals can all provide valuable perspectives on the information used to formulate the objectives as well as a critical review of the proposals. Developing a dialogue with these personnel promotes their support for the agreed-upon objectives by providing the opportunity for input.

An organization that relies only upon historical experience will formulate objectives based upon historical failures. This approach may miss issues that are not obvious from casual observation. As the organization reviews proposed objectives, it may ask questions such as:

- Do objectives address performance for those elements dealing with the highest priority risks?
- Do objectives address potential regulatory non-compliance?
- Do objectives address the identified hazards in operations? Have all the hazards been identified?

- Do objectives address known and perceived weaknesses in the process safety system?
- Have objectives been reviewed and endorsed by all stakeholders?
- Do objectives address improvement trends identified by industry?

If these questions do not have satisfying answers, the organization may wish to modify its objectives. These examples describe different vantage points to use in assessing objectives for the process safety system and should play a role in defining the metrics system objectives.

4.4 DEVELOP THE METRICS STRATEGY FOR IMPROVING THE PROCESS SAFETY SYSTEM

Once the objectives for the process safety system are understood, the next step is to develop a strategy for reaching those objectives. Many actions will likely be proposed to improve performance, but without an unbiased measure to evaluate system performance, the improvement strategies may or may not be effective. This is where the indicators of performance or metrics play a major role in the performance improvement effort. Monitoring the performance of the system elements will provide the important feedback for understanding whether performance improvement plans are effective or may need modification. Metrics are the fundamental data that will help those implementing improvement efforts understand how well these efforts are progressing.

4.4.1 Define the Areas and Elements Needing Improvement

Use the information developed during the objective-setting exercise, as described above, to identify priorities for process safety performance improvement. Select the priorities for improvement within the process safety system elements, and develop a metrics strategy to monitor change for those priorities. For example, if completing PHA studies on time is a priority, setting up a system to track due dates and completion dates for the PHA studies may be selected as metrics to indicate performance. However, if the improvement objective is to monitor the effectiveness of PHA studies, the PHA team make-up, use of the proper PHA method, and the number of previously unknown hazards identified during studies may be selected as the metrics. Data collection related to evaluating the right performance is a key part of the strategy, and detailed metrics will provide the needed information to monitor progress of the improvement efforts.

4.4.2 Define Individual Responsibilities to Support Improvement Efforts

Once the overall improvement plan objectives are identified, detailed strategies need to be defined to successfully accomplish the plan. Strategies need to be translated into specific actions so the parties responsible for collecting and analyzing metrics understand their assigned responsibilities and tasks for implementing the metrics system. Those who collect the data must be given detailed and actionable objectives that translate into activities to support the metrics system. If MOC metrics will be collected, the personnel responsible for conducting MOC reviews will need to understand the data that is to be collected and how that data will be accumulated. These specifics should be consistent across the organization. Defining the detailed objectives within the metrics plan is important to ensure a consistent and successful improvement effort.

4.4.3 Define Performance Indicators to Track Improvement

Defining the indicators that track the process safety system performance is an important structural part of the improvement program. Metrics are used to monitor performance and progress over time. Without a method to track differences in performance, the improvement effort may lack clarity of focus that leads to decreased effectiveness.

4.4.3.1 Select the Appropriate Metrics to Monitor Performance

Selection of the metrics is critical to the success of the process safety system improvement effort. The metrics must provide valid information to support successful changes to the process safety system. Tracking a metric that does not accurately reflect performance will not provide useful information and be of little value. Poorly selected metrics may cause the organization to focus upon the wrong process safety components and to believe, erroneously, that its process safety system is performing quite well.

Such data can create an unwarranted sense of overconfidence and complacency. Linking the metrics to the desired system performance attributes requires serious consideration to gain the most benefit from the effort. There are many resources available to aid in such efforts, including professional organizations such as American Institute of Chemical Engineers (AIChE), trade associations with a

> *Metrics must relate directly to process safety objectives and provide accurate performance information.*

common goal to improve process safety, and many capable consultants that can assist in selection of appropriate metrics and structuring a data collection effort. The CCPS leading and lagging metrics brochure (CCPS, 2007b) identifies a number of lagging and leading indicators that represent a consensus within the chemical and refining process industries on appropriate process safety metrics.

4.4.3.2 Link Metrics to System Goals and Existing Performance

When selecting metrics, the performance improvement objectives and assessment of current performance will play a major role. The improvement objectives set the context within which the metrics are to be selected. Metrics that do not relate to those objectives will not support progress toward the objectives. For example, if an objective is to improve process safety training, metrics that indicate completeness and effectiveness for the training efforts will be selected. Weaknesses that are identified during evaluation of the process safety system will influence metrics selection. Often system weaknesses are the basis for defining some objectives. Recognizing areas where maintaining performance and avoiding performance decline will also factor into selecting metrics. Certain metrics will be tracked where there is urgency to improving the performance of specific system elements. Recognizing where improvement is needed often dictates which metrics should be used to monitor performance improvement progress.

For example, if PHA performance is of interest, the enterprise may choose to track PHA studies by business, by site, and by unit and include data such as:

- The schedule for each PHA
- The due date for each PHA
- The latest completion/renewal date for each PHA
- The PHA team leader
- The PHA team members

Such data determines whether the PHA system is meeting expectations and whether qualified personnel participated in the PHA. This data can be used to identify overdue PHA studies and determine if proper resources are involved in PHA studies. Some organizations document the recommendations from PHA studies, listed by site, including resolution status for each recommendation.

4.4.3.3 Metrics Need to Reflect Process Safety System Performance

Metrics should relate to the elements of the process safety system. Poorly selected metrics that do not specifically relate to the execution of process safety elements will not provide an accurate evaluation of process safety system performance. Occupational injury and illness reporting rates are sometimes used to judge overall safety performance, and this metric does track the incidence of employee injuries quite well. However, this rate does *not* reflect the effectiveness of the process safety system. Occupational safety is quite important to the health and well-being of employees, but the metrics involved in assessing the occupational safety performance are not appropriate for process safety system evaluation; the detailed elements of a process safety program differ markedly from an occupational safety program (as discussed in Chapter 3).

4.4.4   Use Objective Methods to Select Metrics

A technically sound and unbiased method should be used to decide which process safety data to collect. Selection of metrics needs to be based on a representative performance across the organization and not just the very good or very bad performers. Metrics chosen solely using long-held assumptions may or may not truly indicate process safety system performance. If historical-reporting thresholds for incidents are set too high, many lower severity incidents will not be reported. Focusing only on a small number of reported incidents may lead to a sense of complacency and the failure to address issues that could contribute to a future catastrophic incident. Sometimes the selection of metrics might be influenced by available information that is not relevant to process safety performance. All information used to select metrics should have a definable link to process safety performance.

The quality and mechanical reliability areas have used statistical methodologies to design and implement their successful improvement programs. Using statistically valid data and analysis provides a rational assessment of a system's performance. Applying these techniques to process safety systems will support an efficient and effective improvement effort. Statistically valid data will also avoid overreaction to short-term performance anomalies that may not accurately reflect performance of the process safety system; such data will help identify the actual performance improvement opportunities.

Six Sigma[8] and Statistical Quality Control[9] are two examples of statistical data analysis methodologies that have been successfully applied in the processing industries to improve productivity and product quality. Improving process safety system performance is not unlike improving the operating performance. The key is to define individual performance indicators for system elements and procedures, and then develop strategies to increase the reliability of these elements. Using these techniques with process safety data can lead to improved process safety system performance.

4.4.5 Inappropriate Metrics Can Be Detrimental

Metrics should be relevant to the process safety system and accurately reflect system performance. Poorly selected metrics can create an erroneous picture of system performance, which can cause an unwarranted or false sense of confidence in the process safety system if the metrics indicate an unrealistically strong performance. Implementing a poorly designed metrics system will likely waste resources and may lead to poor decisions, generating frustration among those implementing the system and lowering confidence in all metrics data.

Initially selected metrics may not prove to be valuable or appropriate. A useful strategy to validate metrics is to pilot specific metrics at a few, selected locations

---

[8] Six Sigma is a business management strategy, originally developed by Motorola, to identify and remove the causes of defects and errors in manufacturing and business processes. Originally developed as a set of practices designed to improve manufacturing processes and eliminate defects, its application has been extended to other types of business processes.

[9] Statistical Quality Control (SQC) or Statistical Process Control (SPC) is an effective method of monitoring a process through control charts that enable the use of objective criteria for distinguishing background variation from events of significance based on statistical techniques.

that can be trusted to provide accurate information. The pilot results will establish the basis for applying these metrics across an organization.

4.4.6   Set Proper Expectations for Metrics Results

The initial results reported from a metrics system will often indicate performance below expectations. Initial data may be skewed until metrics data collection is reliably implemented, and performance may continue to drop as improvement opportunities are defined and implemented. This sometimes occurs when the historical perspective of performance is too optimistic because objective measures had not been used in the past. Implementing near-miss data collection is one example: The number of reported near misses will probably increase as reliability of reporting improves, so the number of reported near misses is likely to climb during the initial implementation phases of near-miss data collection.[10] Audiences of metrics reports should be prepared for such situations, so the result is expected and potential negative reactions are managed.

4.4.7   Design the Metrics Collection Effort for Available Resources

A performance improvement effort will require time and resources for data collection and analysis. If the metrics data collection effort is too large for available resources, it is likely that specific data will not be captured and that could impact the validity of the system performance evaluations. Any data collection effort needs appropriate resources to execute the designed plan effectively. Sometimes insufficient or incomplete data can be worse than having no data if the reported data leads to false conclusions.

Understanding the resource demand for data collection is critical to successfully designing the improvement effort to obtain valid data. Equally important is to provide the resources for data analysis. Creating a large inventory of data that is not used will not only waste resources but will become frustrating for those who expect to see results that never materialize. It is better to collect less of the most critical data and complete a proper analysis than it is to collect a lot of data and never use it. A critical part of the improvement effort design is to define the data collection and analysis resource needs.

4.5   SELECT METRICS

Earlier sections of this chapter discussed factors that will affect the selection of metrics data to collect, and there are various approaches to select the metrics. Usually the metrics designer assembles a large list of candidate metrics using input from individuals with detailed knowledge of the process safety issues. This preliminary metrics list is usually large and must be pared down to match the

---

[10] See work by Phimister et al. regarding performance of successful near-miss reporting programs.

anticipated metrics system capacity for data collection and analysis. Metrics that are suitable for one facility may not be appropriate for another. Some organizations opt to have a base set of metrics for organization-wide reporting as well as optional metrics that a facility may find useful for its local purposes. The objectives for the process safety improvement effort will also heavily influence metrics selection. There is no single or best approach for selecting metrics since every organization will have unique strengths and needs. However, using the knowledge and experience of others can facilitate the metrics selection.

4.5.1   Choose Metrics Appropriate for the Operation

Consider the operations that will be monitored when selecting metrics. Selected metrics should reflect the most likely problem areas that may arise for the specific operating situation and the available staff to maintain or monitor metrics. Techniques are described and recommended in various publications, such as the U.K. Health and Safety Executive (HSE) publication HSG254 (see Section 4.5.3). Three hypothetical examples are described below:

- *Case 1*—Large continuous operation facilities (e.g., refinery or petrochemical plant) with a high level of automation and personnel. The focus will be on ensuring mechanical integrity and reliability of the processing and process control systems, since automatic systems provide a large amount of process safety protection. Example metrics categories may include:
  - Metrics on process safety events and near misses that have occurred
  - Metrics on status of mechanical integrity system
  - Metrics on adequacy of hazards and risk analyses
  - Metrics for reliability of automation
  - Percentage of critical instruments or safety instrumented systems calibrated on time (e.g., within the timelines suggested by the assessment to meet the requirements of IEC (International Electrotechnical Commission Standard) 61511)
  - Percentage of critical instrument systems found to be defective or miscalibrated
  - Number of critical interlocks that have been bypassed
- *Case 2*—Large continuous operation facility with less automation and a greater reliance upon administrative controls. The focus upon mechanical integrity remains; however, there is a stronger emphasis upon personnel reliably carrying out the manual tasks involved with the operations. Example metrics categories often include:
  - Metrics on process safety events and near misses that have occurred
  - Metrics on timely personnel training and competency demonstration
  - Metrics on status of mechanical integrity system

- Metrics on adequacy of hazards and risk analyses
- Metrics on MOCs, including number of MOCs, status of recommendations, completed training after change, incidents attributed to change, etc.

- *Case 3*—Smaller batch operation facility with lower plant staffing. Example metrics categories often include:
  - Metrics on process safety events and near misses that have occurred
  - Metrics on adequacy of hazards and risk analyses
  - Metrics on MOCs, including number of MOCs, completion of training after change, incidents attributed to change, etc.
  - Metrics on timely personnel training and competency demonstration including abnormal batch procedures

4.5.2   Use Published Guidance on Metrics

Many organizations provide guidance on metrics for evaluating process safety performance. Trade organizations (such as the American Petroleum Institute, the American Chemistry Council, and the National Petrochemical Refiners Association), professional organizations (such as the American Institute of Chemical Engineers and the Center for Chemical Process Safety), academic institutions (such as Texas A&M), and governmental organizations sponsor conferences and symposia that focus on process safety and on other topics such as metrics. The papers and proceedings from these conferences are a resource. The CCPS publication "Process Safety Lagging and Leading Metrics" defines various metrics based upon consensus in the processing and allied industries and is included the appendices of this book. Appendix I lists a composite tabulation of recommended metrics based upon the CCPS's *Guidelines for Risk Based Process Safety* that was augmented by contributions from this book's project team.

4.5.3   Use the Control Barriers Concept

One documented method uses process safety barriers identification for metrics selection. This concept uses a combination of lagging and leading indicators associated with process safety barriers and incident escalation controls to evaluate the process safety system performance. The basis for this method is documented in the U.K. Health and Safety Executive (HSE) publication HSG254 and illustrated by Figures 4.1–4.3. The strength of this technique arises from using the combination of indicators that provides multiple perspectives for judging the surety of a barrier or escalation control. For example, this basic concept was adopted and modified by BP to focus upon three information sources to assess key control barriers as summarized below:

1. Use *hazard analysis* findings to identify potential high-impact events and the
    process safety barriers intended to prevent such incidents. Select metrics that
    indicate the health of these barriers. This is a direct recommendation in
    HSG254, and the BP plant in Hull, England, is piloting this approach (see
    Appendix II). This perspective is seen as a leading indicator.

Figure 4.1 Hazard Analysis to Identify Leading Indicators (BP, 2008)

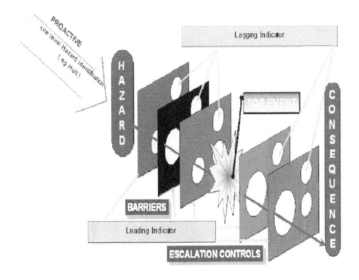

2. Use *incident investigation and analysis* findings to identify process safety
    barrier failures that contributed to incidents. The metrics are designed to
    indicate when the likelihood of repeat failures is increasing. This perspective
    is not suggested in HSG254, but is used by BP at Group Level, based on a
    desire to prevent repeating high-impact incidents.

Figure 4.2 Incident Investigation to Identify Lagging Indicators (BP, 2008)

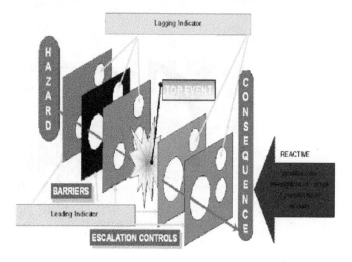

3. Use *shared external learnings* to determine what others have successfully used. Information sources mentioned in Section 4.5.2 are excellent resources for information, such as incident investigation learnings and successful use of various metrics.

Figure 4.3 Learnings From Others to Identify Leading and Lagging Indicators (BP, 2008)

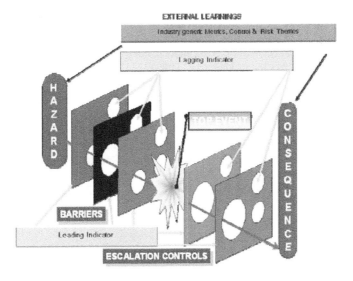

4.6 CONCLUSION

Setting the process safety metrics system goals and objectives is a very important activity since it sets the basis for designing and operating the metrics system. Goals comprise the long-term vision of the desired future state, and objectives define incremental progress toward those goals. Senior leadership is needed to initiate the development of metrics goals as this demonstrates their commitment, and senior management must provide funding and resources to demonstrate their long-term support if the metrics system is to achieve its objectives. Metrics are selected to support reaching objectives, and there are various ways to choose the appropriate metrics. Evaluating current process safety elements performance (relative to expectations), existing process safety risks, and/or recognized strengths and weaknesses of existing systems are a few of the techniques that can be used. Considering multiple views such as operating personnel, process safety professionals, and engineers can shape a complete picture of process safety performance that will aid in metrics selection. It is important to plan tracking progress toward meeting objectives (and goals), and to periodically evaluate individual metrics to ensure that the information being collected is bringing the anticipated value. A successful metrics system is a long-term endeavor that can help improve process safety performance. Selecting appropriate metrics using unbiased and broad-based information will lead to a high-performing system.

REFERENCES

California Accidental Release Program (CalARP), California Code of Regulations, Title 19, Division 2, Chapter 4.5

Center for Chemical Process Safety, "Process Safety Lagging and Leading Metrics," AIChE Industry Technology Alliance, December 20, 2007

Contra Costa County (California) Industrial Safety Ordinance No. 98-48, now Chapter 450-8 of the County Ordinance Code

Control of Major Accident Hazards (U.K.), S.I. 1999/743, implementing the Seveso II Directive (96/82/EC) in Great Britain

Delaware Risk Management Plan, Chapter 77, Title 7, Delaware Code

Health and Safety Executive (HSE), "Step-By-Step Guide to Developing Process Safety Performance Indicators, HSG254," Sudbury, Suffolk, UK, 2006.

Nevada Chemical Accident Prevention Program (CAPP) (NRS 459.380)

Phimister, J. et al., "Near-Miss Incident Management in the Chemical Process Industry," *Risk Analysis,* Vol. 23, No. 3, 2003

Toxic Catastrophe Prevention Act (TCPA) N.J.S.A. 13:1K-19 et seq.

5

IMPLEMENTING A
METRICS PROGRAM

Implementing a metrics program requires strong management support and leadership throughout execution and a well-thought-out, detailed plan. Organizations that rush to implement a metrics program will likely stumble at a later date due to a lack of sustained management support, poor implementation planning, a lack of resources, a lack of understanding of the data, or a perceived burden by those asked to provide data. Expending the effort to develop an implementation strategy, focusing on the most meaningful and beneficial metrics first, will pay off in the long term. Metrics that provide little value or are of little interest will hinder the successful implementation of the program. This chapter outlines a process for implementing a metrics program for a department, facility, or organization.

5.1 MANAGEMENT SUPPORT AND LEADERSHIP

Management support and leadership during implementation of the metrics system is essential to success, and management needs to be engaged regularly throughout the implementation. Management's role is to provide overt support and reinforcement for the implementation objectives, and management leadership builds buy-in for the plan by demonstrating commitment and support for the strategy. Regularly engaging management provides them opportunities to identify and resolve roadblocks that may hinder successful system implementation. This also provides management the opportunity to reinforce the reasons for establishing the metrics system as well as allow them to monitor the progress.

5.2 DEVELOP AN IMPLEMENTATION STRATEGY

The goal of a metrics program is to provide objective measures for system performance. This is true for all management—operating or financial—systems. A well-defined implementation strategy will provide a solid foundation and road map for the individuals charged with putting the program in place and sustaining it into the future.

5.2.1   Define the Goals and Objectives of the Implementation Strategy

Once the organization has established which metrics it will implement, the next step is to establish a detailed implementation strategy with specific objectives. Any management system needs a strategy document that defines the objectives, purpose, scope, and methods to be used for meeting the system goals. The objectives for the implementation strategy are different from the discrete metrics objectives discussed earlier in this book. These objectives are related to the implementation of a management system and are associated with the tactical nature of launching the system.

The strategy document will describe scope of coverage for the metrics system, detailing what is included in the system and what is not, and will be used to guide development of detailed procedures for implementation and operation of the metrics system.

5.2.2   Define the Management Strategy for Implementation

There are various ways to manage an implementation effort and the ongoing operations of a metrics system. An often-used strategy follows the widely documented Plan-Do-Check-Act model, which is described in Chapter 2.

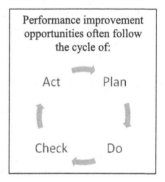

Adopting a defined management strategy at the onset of implementation ensures that all subsequent plans follow a consistent approach in developing strategies and plans and conducting proper management reviews.

5.2.3   Establish Implementation Milestones

When an entity begins defining the implementation strategy, high-level milestones are identified and matched with target completion dates. The implementation steps outlined in this chapter are summarized in a Metrics Implementation Milestone Chart, which illustrates relative timing, in Figure 5.1.

Figure 5.1 Metrics Implementation Milestone Chart

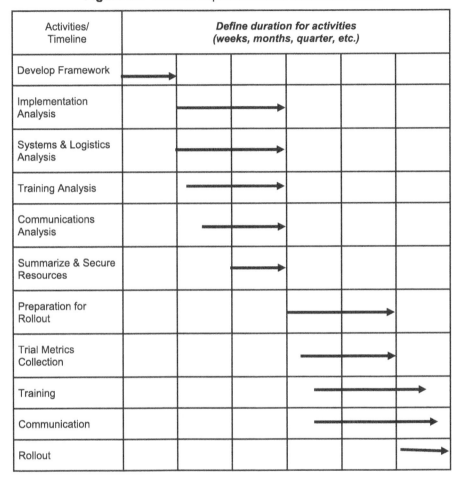

Activities/ Timeline	Define duration for activities (weeks, months, quarter, etc.)					
Develop Framework	⟶					
Implementation Analysis		⟶				
Systems & Logistics Analysis		⟶				
Training Analysis		⟶				
Communications Analysis		⟶				
Summarize & Secure Resources			⟶			
Preparation for Rollout				⟶		
Trial Metrics Collection				⟶		
Training					⟶	
Communication					⟶	
Rollout						⟶

Implementation milestones should be approached thoughtfully. Depending on the breadth and depth of the metrics system, the timing and implementation scope will vary. For example, if a corporation has chosen three key metrics to implement globally, and understands that systems are already in place to capture the new metrics, the defined milestones may be reached more quickly than illustrated in Figure 5.1. Alternatively, if a large facility has chosen a dozen metrics to be implemented in multiples areas, such as in maintenance, operations, and the engineering group, the system may require more coordination and time to fully implement. If fewer resources or systems are available to support collection and analysis of the metrics, then the implementation timing may need to be extended until resources are added or available to complete an effective implementation.

The option exists to scale back the number of metrics collected initially, selecting various metrics to collect first and phasing in the others over time.

Organizations should look to their management for guidance regarding implementation milestones during the early planning stages for the metrics system. Management may desire a definite timing for full implementation of the metrics system or may focus on a specific date for reporting the first set of results. It is important to understand the expectations of all stakeholders when defining the objectives of the implementation strategy.

5.3 DEVELOP THE FRAMEWORK FOR THE METRICS IMPLEMENTATION STRATEGY

Successful and effective implementation of a process safety metrics system is done by first defining the structure for the system. The structure or framework includes how data is collected, analyzed, and shared with the organization. The framework defined in this section will add more detail to the overall objectives and goals earlier in the chapter and will provide a foundation for the analysis conducted in the next section of this chapter.

5.3.1 Implementation Scope

Those who design the process safety metrics system must carefully define the scope or coverage of the system. If the scope is not adequately defined along with the responsibilities of personnel who will collect and use the data, then confusion and loss of focus within the metrics system implementation can occur. In addition, the results represented by the metrics may be misunderstood, if what is and what is not being measured are not defined well and understood. For example, a facility maintenance department may develop metrics for completion of mechanical integrity inspections. Without defining and communicating the scope of this metric, someone reviewing the summarized data might assume that the metrics represent inspection and testing for relief devices and safety instrumented systems, when in reality the maintenance department is only tracking piping and vessel inspections. The scope must be described in detail to ensure it will meet stakeholders' expectations and is understood throughout the organization.

The metrics system will need criteria to determine which units within a facility, or which facilities within the organization, will participate in the metrics system. Many companies, and even some large plant sites, have diverse operations, some of which may not use typical process safety systems, and this usually applies to low-hazard and low-risk operations. These facilities should be included in the process safety metrics system, and the system design needs to establish participation criteria for all. For example:

- Some organizations choose to have every facility participate in their process safety management and metrics systems using a basic set of metrics.

- Some organizations feel it is appropriate that only those processing units that meet defined hazards criteria participate in the process safety metrics systems.

- Some organizations mandate that all units at a site participate, even if only a single unit meets their criteria; they believe it is better to establish a single culture for the entire facility rather than support different process safety requirements for individual units.

The organization needs to clearly define the participation criteria for its facilities to avoid confusion and unproductive dialogue on who must participate. The metrics systems design should embody clear criteria for participation that addresses all anticipated situations within the organization.

Some criteria for determining which parts of the organization will participate in the metrics system may include:

- *Hazardous Materials Threshold Quantities.* Some organizations use threshold quantities (TQ) for hazardous materials to determine if a unit will participate in the metrics system. This is similar to many regulatory methods that use a specific list of hazardous materials coupled with an on-site inventory level to determine if a facility or process is covered by the regulation.

  TQ values normally vary with the hazardous nature of the materials handled. If a small quantity of material will cause a major process safety incident, like an inhalation hazard such as cyanide, the TQ is set low. If a large amount of material is needed to create a dangerous situation, like a flammable liquid, a higher TQ is set.

- *Hazardous Nature of Process.* Some processes have increased hazards and risks due to process operating characteristics. Though the materials being processed may not reach a TQ for a hazardous material as described above, the characteristics of the operating process may present a hazard with the potential to cause a significant process safety incident. Processing characteristics used by some organizations include processes that:
  - Operate at high temperatures and/or pressures;
  - Operate at extremely low temperatures and/or vacuum;
  - Handle combustible dusts;
  - Generate exothermic reactions; or
  - Batch operations with significant material compatibility issues if not sequenced properly.

- *Regulatory Criteria.* Some organizations use regulatory information to determine which facilities will participate in the metrics system. The logic is if a facility is covered by a process safety–focused regulation, then that facility should participate in the metrics system. Examples of such regulations include the U.S. OSHA Product Safety Management (PSM) rule,[11] the U.S. EPA Risk Management Program (RMP) rule,[12] and European regulations arising from the EU Seveso Directives,[13] such as the U.K. Control of Major Accident Hazards (COMAH) regulations.[14] A company may choose to have only the facilities covered by a regulation participate in a metrics system. However, process safety hazards do exist in facilities not covered by regulations, and many companies opt to extend their metrics participation to cover any operations that represents a potential process safety risk or where operations within a unit may affect a facility containing hazardous materials.

• *Company Process Safety Standard.* Some companies define internal requirements and/or standards for implementing process safety systems across all their processing operations that handle hazardous materials and/or high-hazard-processing conditions. These companies require that any unit that meets its internal criteria for implementing its process safety standard will also participate in the metrics systems.

Once the organization defines participation criteria for the metrics system, the organization can document those facilities using a metrics matrix as illustrated in Figure 5.2.

Figure 5.2 Example Metrics Matrix

Location	Process Safety Incident Metric (Count, Rate, Severity)	MOC Metric (# Completed per Month)	% PSM Inspections Completed (YTD On-Time)	% OT Production & Maintenance (Monthly & Qrtly)	% Inventory Weekly Inspections Completed (YTD On-Time)
XYZ Corporation – all locations	X				
XYZ Corporation – New York	X			X	
Batch Chemical		X	X	X	

---

[11] 29 C.F.R. pt. 1910.119.

[12] 40 C.F.R pt. 68.

[13] European Union Council Directive 96/82/EC, as amended by Directive 2003/105/EC.

[14] Statutory Instrument 1999 No. 743.

Location	Process Safety Incident Metric (Count, Rate, Severity)	MOC Metric (# Completed per Month)	% PSM Inspections Completed (YTD On-Time)	% OT Production & Maintenance (Monthly & Qrtly)	% Inventory Weekly Inspections Completed (YTD On-Time)
Mixing Dept.					
Reactor Dept.		X	X	X	
Shipping/Receiving Dept.					X
Maintenance Dept.			X	X	
XYZ Corporation – London	X			X	
Waste Treatment Dept.		X	X		
Reactor Dept.		X	X	X	
Shipping/Receiving Dept.					X
Solvent Recovery Dept.		X	X	X	
Maintenance Dept.			X	X	

5.3.2 Identify Metric Leaders

Organizations that initiate new management systems often assign a "champion" for the effort. The champion plays a central role in developing system strategies and driving development progress for the metrics system. The champion is responsible for coordinating metric system development activities, developing justification for resources, and maintaining communication with management on the status of the effort. Depending on the size of the metrics system and how far throughout the organization the system will be implemented, champions may be needed for different levels or functions. Some organizations choose to assign the corporate champion role to a high-level management leader or a member from the board of directors.

5.3.3   Stakeholder Understanding and Commitment

Successful implementation of the metrics system requires commitment of the site and unit managers as well as those who supervise personnel directly responsible for collecting the data. Managers and supervisors will play a key role in the success of the metrics system, and it is important for them to understand what the metrics measure, why the metrics are valuable, and how they and the company will benefit. As managers and supervisor gain a full understanding of what the metrics mean, they can take appropriate actions and decisions, and can reinforce the commitment to fully complete the metrics system procedures. The managers and supervisor will have the responsibility to ensure personnel understand their role in the metrics system, review metrics results, and help formulate actions based upon those results.

> *Aspects to consider when designing a data collection system:*
>
> - *Leveraging existing data collection tasks*
>
> - *Creation of a new data collection task(s) for an individual/group*
>
> - *Programming requirements for modifications to existing systems*
>
> - *Database or spreadsheet requirements for new data collection systems*
>
> - *Frequency of data input*
>
> - *Data outputs and reporting features*
>
> - *Accessibility and security requirements for data input and reporting*
>
> - *Data validation or auditing*

The benefits of a metrics program should be emphasized when soliciting the understanding and commitment of stakeholders. Some may push back or resist another program, likening it to "the program of the month" or another requirement from corporate. Process safety metrics are not a hindrance to the business, but a means of tracking progress towards the overall goal. The benefit of reaching the ultimate goal of zero process safety incidents is untold savings of life, environmental impact, property, and business. In addition, there are indirect benefits of tracking progress towards the goal. For example, if a facility implemented a metric regarding quality and quantity of operator training, the benefit of improved performance is a better-trained operator. The more competent operator not only will operate the process more safely, but may also operate the process more efficiently and improve quality. Downtime may be reduced due to decreased errors and increased intervention to mitigate abnormal situations.

Involving stakeholders in the development of the implementation framework and soliciting their buy-in will help minimize surprises during actual implementation. Even though details and specifics are not necessarily outlined at this stage in development of the implementation framework, conveying the vision and a high level of information can be helpful in gaining understanding and commitment. In doing so, the stakeholders may offer support, information, or other systems that can be useful when implementing the management system.

5.4   IMPLEMENTATION ANALYSIS

With the framework of the implementation plan established, the details of the plan can be developed. This is done by reviewing and analyzing several aspects prior to preparing for the rollout. If the collection of metrics is difficult or the necessary tools are not available, it is likely that the metrics system's performance will not meet expectations. During the metrics system development, the designer(s) should identify and account for potential barriers to reliable and routine system operation and consider how data will be collected, validated, stored, analyzed, and communicated. Any barriers to routinely completing metrics procedures will be detrimental to reliable execution of the metrics system; taking the time for this analysis prior to implementation can lead to a more efficient and effective metrics program.

5.4.1   Systems and Logistics Analysis

Procedures and activities associated with the collection of metrics are best integrated into the normal work of the personnel collecting the data. Any special procedures that are viewed as extra or unrelated work may not be endorsed by personnel charged with the data collection. If an individual is focused solely on the metrics program, data collection and analysis tasks are defined as part of his or her job. However, much of the data will be collected by personnel whose major responsibilities are not directly associated with the metrics system.

In analyzing the envisioned metrics data system, several questions should be answered regarding the data and logistics associated with collecting that data. For example:

- Is the definition and scope of the metric understood?
- What data is needed to develop this metric?
- Is there a guidance document to describe the essentials for each area of a measure?
- Is it understood where and how to gather the data for the metric?
- How is the data to be collected (e.g., automated or manual entry)?
- Is a given metric already being tracked by another function or data collection system?

- How is the data validated or audited (e.g., control system voting, redundant entry systems, personnel assigned to check the data)?
- How often and when is the data collected and reported?
- What does the reported data look like? Is it in a format that is ready to import into the metrics program or will it need to be further analyzed with additional tasks prior to importing it into the metrics program?
- Who has access to and how secure is the data?

Using the list of existing data collections as described in Section 5.3.1, the organization can begin to analyze how best to integrate current data systems and what information gaps will need to be filled by additional or alternate means. Working with the data collection system owners to document the answers to these questions will often provide sufficient detail about the existing system and collection of data. This will identify information needs or system modifications necessary to implement the metrics system.

Once the information needs and system modifications have been identified, work should continue to fully design data tasks for the metrics system. Avoid designing data collection tasks that require a significant departure from the responsible individual's normal activities. Requiring an individual to redirect his or her attention to a seemingly unrelated task can cause an individual to overlook or defer a metrics task due to the demands of his or her primary responsibilities. Designing the data collection tasks as an easy extension of existing responsibilities will achieve better acceptance and more reliable performance.

It is important to collect data that will provide valuable insights into system performance. Sometimes, developers of the data systems start with all existing data, but do not have the knowledge to cull out the less valuable information. This leads to collecting information that is not very valuable and consumes resources with little gained in return. Metric system implementation leaders need to carefully select the final metrics that provide valid representations of performance.

5.4.2   Leverage Existing Data

Information and data collected by other organizational data systems might be valuable to process safety efforts, but if other databases are fed into the metrics system, the interface between these systems must be defined. Organizations should identify what data collection systems exist and list those in the framework strategy.

For instance, many objectives and data collection similarities exist between the mechanical integrity (MI) system and the process safety metrics system, and data available in the MI system can be used to judge process safety performance. How information from the MI system will be captured and used needs to be defined, including who will be responsible for ensuring the proper information flow. This requires the metrics system developers to coordinate with the MI personnel regarding what information will be monitored and how it will be transferred to the metrics system. The metrics system will likely monitor "inspections completed per schedule," and it will be necessary to select which inspections to monitor. Often, recognized and generally accepted good engineering practices (RAGAGEP) for process safety inspection programs, such as the American Petroleum Institute (API) standards 510, 570, and 653,[15] are used to track process safety performance. Inspections that are not related to process safety, such as inspecting administrative building elevators, are not monitored by the metrics system.

> *Potential data systems to leverage:*
>
> - *MI inspection and testing*
>
> - *Process control historians (i.e., challenges to safety systems)*
>
> - *Environmental release and loss of containment events*

Many data collection tasks can be automated with the more powerful electronic control and instrumentation systems that are common in processing operations. Processing upsets and other processing data can be automatically recorded using these systems in "process control historians." For example, challenges to safe operating limits are of interest to operating management who wish to eliminate operating excursions. This information is also of interest to those charged with process safety responsibilities who need to decide if adequate layers of protection are in place to safely handle these recorded challenges. This data may already be collected in the process control system historian and can be used as an input to metric. For example, with some minor modifications to the historian reporting features, the system could be programmed to print out a daily report of challenges to the safe operating limits along with the other daily production reports. This information can be transferred to process safety personnel and used for discussions during shift meetings. Further, the historian reporting could be programmed to print out a monthly log of challenges to safe operating limits to roll up into a monthly metric. With some minor programming changes, this data-reporting task has then been incorporated in a near transparent way into the daily tasks of production personnel. Automating the data collection process can provide reliable data with little increased demand for human resources. The details of the data collection design can be documented as part of the implementation strategy.

When considering existing data collection systems that may provide information for the process safety metrics system, several considerations should be evaluated. A few questions to illustrate these considerations include:

---

[15] API Standard 510: Pressure Vessel Inspection; API Standard 570: Piping Inspection; API Standard 653: Tank Inspection, Repair, Alteration, and Reconstruction.

- Does the existing data collection system have all the information required for the process safety metric? If not, is the data collection system capable of collecting the required information?

- Does the existing data system operate on a frequency that is consistent with the proposed process safety metrics needs? If not, is the existing data collection system capable of collecting information at the frequency required for the process safety metric?

- Can the candidate existing system be modified to include objectives for process safety metrics? Are modifications practical?

- Could the existing system output be substituted for the initially proposed process safety metric?

5.4.3   Data Validation

When implementing a metrics system, it is important to ensure the process safety data is reviewed for accuracy. Inaccurate data can lead to poor decisions and focus improper priority to issues. Worse, inaccurate data may focus attention away from serious performance deficiencies. The metrics system designer needs to define the methods that will be used to validate data entered into the metrics system. There are several techniques for validating

> *Poor quality data leads to poor decisions.*
>
> *Adopt a philosophy of "trust but verify."*

data; many of the techniques have been developed through quality-based efforts and auditing methods. The following is not a detailed "how to" for developing a validation method, but rather introduces topics for further research.

5.4.3.1   Plan to Validate Data

The implementation plan for the metrics system should include a component that confirms the accuracy of the collected data. A broad-based metrics system could cover multiple sites across diverse geographical regions, and there is opportunity for misunderstanding as to detailed definitions for the data collected, the frequency of reporting data, and other specifics of the metrics system. Formally reviewing the process safety data can identify any data problems such as improper data being reported or an identifiable lack of data entries. It is not uncommon for new data systems to experience problems that can stem from inefficient, poorly defined data-reporting methods and communications that fail to set the proper priority for data reporting. Data validation will help determine where problems exist, allowing for corrective action. For example, assigned data reporters may not feel data entry is an important priority and may neglect entering data due to other work assignment priorities. The implementation plan needs to identify weaknesses in the execution of the metrics program. Because collecting and reporting inaccurate data can impact the metrics system success and lead to erroneous conclusions, validating data in the implementation plan is a critical element.

5.4.3.2 Conduct Pilot Studies to Validate Value of Metrics

Initially selected metrics may not prove to be valuable or appropriate, and it is often difficult to predict what metrics will work well for intended purposes. One useful strategy is piloting specific metrics at a few, selected locations to determine if the metrics provide expected and valuable data. Selecting the sites at which to pilot metrics should be based upon a site's capability to conduct a pilot study and to provide accurate information. The pilot results will establish if the piloted metrics are worthwhile to deploy across the organization.

- *Statistical analysis techniques can validate data.* Many proven techniques exist to aid in validating data entered into a metrics system. Some statistical analysis methods are applicable for large data systems while other, simpler techniques can be effective for smaller data systems. The sophistication and automation of data analysis is usually dictated by the size of the data system and the cost savings associated with increased efficiency for other resources. Statistical quality control methods show unusual trends and anomalies in data entries. The use of control charts is one technique to analyze data; control charts plot the data and establish "normal" values and "control limits" for specific periods of time. Anomalies in reported data can be identified and evaluated to understand if the change is real or a data-reporting problem. Examples of statistical data analysis techniques are also found in "computerized maintenance systems" that analyze large amounts of data, tease out trends, and highlight systematic failures. Unexpected results can be evaluated to confirm if data reports are valid.

- *Auditing methods can validate data.* Audit techniques are commonly used to validate data entries into a metrics system. Such techniques may include reviewing the data and asking some selected questions, such as:
  - Was the data reported properly according to the predetermined definitions?
  - Was all data reported reliably? Are there any gaps in data entries?
  - Is the system in place to capture the desired data? Do all data reporters understand their role?
  - Is there a process in place for resolving questions about data reporting or to highlight data anomalies?
  - Does the data indicate expected performance? If not, can this be explained?

Audits are powerful tools for understanding if data entries are completed as intended. Reported data reviews with those who entered the data can quickly determine if the data was properly reported and the correct definitions were used.

Often auditing selected partial data sets can indicate if the data system is in good shape or if there is potentially a systemic problem. It is also possible to determine if problems exist with the data collection system or with execution of system-defined tasks.

Auditing the process safety metrics system is a key component to ensure that the program is being implemented as planned. As the implementation plan is developed, auditing the processes of data collection, data analysis, metric reporting, etc. should be included as specific steps. Checking these steps early in the implementation phase will allow deficiencies to be identified and corrections made. Once the implementation is mature, periodic checks of these processes will ensure that the designed metrics program is maintained. However, process safety audit teams must be specifically tasked with auditing the metrics system. If the scope of the process safety audit does not specify validation of the metrics program, asking the team to do so may divert the primary focus of the team's efforts. In the case of compliance audits, process safety audit teams should rely on metric results as a trusted input into the audit to help focus their attention to underlying issues of underperformance.

5.4.4   Resource Analysis

When beginning to develop a metrics program, the tendency is often to collect all information that may relate to the objectives of the system. However, the scope of the metrics system should balance the effort to collect and analyze the data with the value that the data will provide. Collecting data that will not reflect important indications of process safety or process design performance brings little value. Collecting too much data may overload the resources available to reliably collect and analyze the data. Collecting unnecessary or unused data can be counterproductive as a performance improvement effort and waste scarce resources; it can also create unnecessary legal liability since the data could be subject to differing interpretations by other parties.

Overburdening the metrics system (and those who use the metrics) with too much data may obscure important trends and open the possibility of focusing on the wrong issues. Developers of metrics systems need to be aware of the resources required to routinely collect and analyze the data. Specifying too much data collection will overburden the available resources for data collection and analysis. Any effort to collect data will compete with objectives in other areas.

When analyzing the resources needed for implementing a metrics system, there are several considerations once the data collection systems have been designed. The first consideration is to determine if the resource needs are above the current available resources. For example, if data for the mechanical integrity metric will be supplied by existing data, from existing reports, and from existing tasks, any additional time requirement is small. However, if the data supplied for the mechanical integrity metric will need to be organized and analyzed by maintenance personnel before submittal, there may be an additional resource need. Aspects to consider when analyzing resources may include:

- Estimating the number of hours and with what frequency the task will be completed (e.g., 2 hours per month, 20 minutes per week, 5 days per year)
- Determining which individuals or roles will have the responsibility for completing the data collection, analysis, and communication tasks
- Identifying a back-up resource to assist with data collection in the event of absence, illness, holiday, etc.
- Understanding the computer programming and/or database requirements and estimating the IT-related hours required to implement the new metrics
- Reviewing the data collection design and estimating the specific resources required to complete the data collections tasks.
- Determining whether there are human resources or whether revised (or new) computer applications are necessary to ensure sufficient resources to complete the data collection.

If the resources are insufficient to carry out the assigned system tasks, reliable execution of the metrics system is unlikely and consideration should be given to either increasing the resource commitment or potentially scaling back the scope of the system. It is usually better to operate a smaller system that provides reliable and accurate results than to continue operating a larger system with inadequate performance. However, prior to reducing the scope of a metric, it is important to evaluate the impact on the metric if the scope is reduced. If reducing the scope of the metric results in less-meaningful information, the reduced-scope metric may no longer provide valuable information. Asking for an "outside the system" review by an unbiased party may provide insights for assessing performance in operating the metrics system.

5.4.5   Communication Analysis

During implementation of the metrics system, efficient and appropriate communications with stakeholders are important. A well-thought-out communications plan ensures that personnel have the information they need to fulfill their duties. This is a two-way activity since implementation leaders need up-to-date information to appropriately manage the implementation. During implementation, periodic updates for management keep them informed and provide opportunities for management to show support and facilitate implementation. The communication analysis is an activity to determine which audience needs what information during implementation and how often to communicate.

When developing the implementation strategy, it is also important to understand what metrics need to be reported to what audiences. Chapter 6 discusses communication strategies for several important audience groups.

5.4.6   Training Analysis

Attention must be paid to how the system will be operated, including the education and training of those who will implement the system procedures. Documented training systems, such as the Systematic Approach to Training (SAT) that is used successfully by the nuclear power industry and the Department of Energy, can be helpful in developing training plans (DOE, 1994). Many existing third-party training resources can also aid in defining a training system. When collecting data over a diverse set of facilities and using this data to assess overall process safety system performance, it is necessary to ensure that all parties use the same metrics definitions. It is not possible to make quality assessments if differing definitions (or interpretations) are used by different data contributors. Periodically confirming the use of consistent definitions by all parties will build confidence in the integrity of the data.

Complicated data systems require time for educating and training personnel who will use, administer, and evaluate the metrics systems. Therefore, the metrics system developer should define education and training needs of those who will complete the metrics tasks as well as the required training resources and include this in the implementation strategy. (Training is discussed in more detail below.) Easier-to-operate metrics system may require less training than a more complicated system requiring significantly more training.

Education in the following areas will be necessary for those expected to carry out data collection and analysis tasks:

- Objectives and goals of the metrics program
- Objectives and goals of discrete metrics
- Requirements for individual tasks, such as data collection, analysis, and reporting
- Frequency of expected task completion
- Resource requirements for the tasks
- Requirements or strategies for auditing or validating data
- Resources available for program support or to answer questions

If proper training is not accomplished, smooth implementation of the system is unlikely, valuable time and information will potentially be lost, and time and resources will be wasted due to reworking and correcting incorrect data. The worst outcome is that invalid data may be used to make improper decisions.

Develop the detailed objectives of the metrics training system and define what outcomes are expected from the training and how these outcomes will impact the implementation of the metrics system. There will be broad principles that need to apply across the organization and specific task-related knowledge that must be transferred to the practitioners for the metrics system. As the training is defined, estimates for resource needs and time to complete the training can be made. It will be critical to define the training resource requirements as the metrics system is developed so resource commitments can be integrated into objectives.

Training activities do not always meet the defined objective of providing sufficient knowledge and information. Training may contain too much material to be absorbed or not contain enough reinforcement of the key components. If data collection follows a frequent routine, it is likely that constant repetition will help support reliable execution. However, if the data collection is sporadic, the likelihood of reliable execution declines. Often follow-up training is carried out to increase reliable execution.

5.5   PREPARE FOR ROLLOUT

When the implementation plan has been developed, the next step is to start implementing the plan by preparing the organization for a full rollout of the metrics program.

5.5.1   Summarize and Secure Resources

Senior leadership is vital to providing implementation momentum (and continued operations) for the metrics system. If management support is not sustained throughout implementation and subsequent operations, implementation success can suffer as resources are shifted to competing activities or other priorities are stressed. Any management system effort requires resources to develop, implement, and support ongoing operations. This support will need to come via a commitment by those who make business and resource allocation decisions for the organization. Small programs may be initiated at the site or unit level where the manager has the authority to assign personnel and approve expenditures for systems under his or her purview. A large system that spans multiple sites or countries will likely need the support of corporate and/or business management.

At this stage in the implementation of the plan, the implementation strategy and analysis should be presented and discussed with management. As with the reporting of metrics, the level of detail of implementation information shared with management may depend on the level of management participating. For example, the unit-level manager may seek a more detailed overview of the resources, systems, and time needed to completely implement the strategy. On the other hand, a more senior corporate manager may only want to know the "bottom line"—how much money is needed to implement and when will results be available. This is a good opportunity to review the plan with appropriate levels of management and secure the resources for effective implementation.

Even if the metrics are collected, without appropriate resources, the data may not be analyzed and reported to management. If full support for the plan is not achieved, the persons charged with implementation may choose to reduce the number or sophistication of metrics initially, gaining the needed support for the program, then revisit the desire for additional metrics at a later date.

5.5.2   Training and Communication

Individuals who collect and analyze metrics—those who will lead the implementation as well as those who will receive and use the metrics—need to understand why the metrics are valuable and how the company, and they personally, will benefit from their adoption. If the personnel involved in collecting the data understand the objectives and benefits of the metrics program, it is more likely they will develop the commitment to execute the metrics system procedures. Understanding value is important but not sufficient, and personnel responsible for collecting information must also be trained on how to collect the information accurately and consistently. Training and education is therefore important to the success of the metrics program.

5.5.2.1   Train Management and Leaders

Management leaders have important roles in a successful metric system implementation. They need to demonstrate commitment to their responsibilities during the implementation (and continued operations) and thoroughly understand the system justification and its value for the organization. Management leaders need to understand metrics' definitions as well as the rationale for selecting specific metrics. Leadership's enthusiastic support and ability to engage in metrics' dialogue will add momentum to the implementation efforts. To meet their responsibilities, leaders need to meet their own objectives and hold their personnel accountable for assigned implementation activities. Often these personnel are trained in metrics system corporate expectations and how to drive implementation success.

5.5.2.2   Train Knowledgeable Practitioners

The metrics system will require informed practitioners to reliably complete the system tasks. There are many examples where poorly trained personnel collected data that was misunderstood or just wrong. This wastes resources by collecting improper and potentially useless data and may lead to erroneous conclusions about performance. A false sense of confidence could result from improper data that indicates the process safety system is operating more reliably than is true. On the other hand, invalid data may lead to a conclusion that action is needed when, in fact, the process safety system is meeting expectations.

5.5.2.3   Educate Audiences on Metrics Objectives

Training should include education on the goals and objectives for the metrics system and what value will be gained from the effort. The audiences should also understand their responsibilities with regard to the data. Not only should they recognize and celebrate success, they should also recognize and take action when the data shows a decline in performance. Just as many organizations recognize or even celebrate when a certain period of time passes without an occupational injury, many others will monitor injury rates and take measures to prevent further injuries when the metrics indicate an increased rate of occurrence. Similarly, teams will often celebrate a project getting completed early and under budget, but will also monitor the calendars and budget closely if timing and costs are getting tight.

Some metrics, such as number of fires or number of process safety incidents, may individually portray the performance of a process safety program. Other metrics may have information that is valuable but requires an understanding of the underlying meaning of the metric. Examples may include:

- A facility decides to measure the percentage of overdue inspections for critical piping due to the number of leaks that have occurred at the facility. As performance improves and the goal of zero percent overdue is achieved, it may be discouraging when the next piping leak occurs. This is an opportunity to coach the audience that there may be a residence time to allow the improvement in the management system to be reflected in improvement in performance. It may also indicate a need to focus on the next objective once the first objective has been achieved. For example, when the metrics indicate MI inspections are being completed on time, the next objective may be to focus on the quality of inspections.

- An organization has been measuring the number of process safety incidents, as defined by corporate guidance, at all global facilities for several years. Recently, the organization initiated an effort to redefine a process safety incident, provide new tools to collect and analyze the data, and relaunch the program globally. Initially, the number of incidents decreases, but after a few months the number of process safety incidents increases dramatically. Is this truly a sign of a negative trend? In some cases, it could be. More likely, the downturn in performance as indicated by the data may be a function of increased reporting and better accuracy in the data.

- A facility measures the number of near-miss and incident investigations findings for corrective actions. As a stand-alone metric, this can help the facility understand the depth of the investigation. Fewer findings could indicate a less-complete investigation. However, the audience receiving this metric could be coached that evaluating the number of near-miss or incident investigations could also provide insight on the performance of the process safety management system. For example, if incomplete (or a lack of) operating procedures are identified as a frequent contributing factor to near misses or incidents, this should signal an opportunity to evaluate and improve the operating procedures management system.

- A facility tracks the number of completed recommended corrective actions from PHAs. This metric alone will convey the progress of completing recommendations. However, the audience receiving this metric could be coached that improving performance of 50 percent or 60 percent, or 99 percent complete is positive, but the 50 percent,

40 percent, or 1 percent of incomplete PHA recommended corrective actions may signal inefficiencies or lack of resources to complete.

Failure to follow up on this information as described in the last two examples could be a contributing factor to a future incident. It may also be an indication of the organization's lack of commitment to improving process safety performance.

By illustrating the benefits of the metrics system and individual roles and responsibilities, commitment to work toward the success of the system can be built among the practitioners. When new systems or major change is implemented, it can be difficult to gain a full commitment from those affected unless they know why the change is being made and the value the change is expected to generate.

5.5.3   Trial Data Collection

Organizations may choose to conduct a trial collection of data. This is a short-term, time-bound test of the metrics program. The purpose of conducting a trial is to identify roadblocks to successful implementation of the entire program, possibly including:

- Are the resource requirements and estimated time to complete the metrics tasks accurate?
- Are additional resources needed to complete the tasks?
- Are the metrics meaningful with respect to the organization's goals?
- Is the frequency of data collection appropriate and adequate?
- Is the communication of the metrics meaningful and understood by metric stakeholders?
- Are the validation and auditing process effective and efficient?

Improvement opportunities identified during a trial period should be resolved prior to a full rollout of the program. Adjustments to some details of the implementation plan can be expected. The experiences of conducting a trial rollout will help organizations gain confidence in the program and may also increase the likelihood of implementing an effective and efficient metrics program.

5.6   ROLLOUT

When the facility or organization is ready to begin fully implementing the metrics plan, the time and energy spent in developing the strategy will help ensure an efficient and effective implementation.

5.6.1   Collect Data

With training completed, communication of the program established, and implementation of data collection systems, data is ready to be collected. The persons tasked with this responsibility should already understand when to start collecting data and how to manage the data once collected.

5.6.2   Management Review of Metric Data

Management has played a key role in the development and implementation of the metrics management system. It is imperative that management continues to be engaged through the initial rollout period and on an on-going basis. Their review of initially collected metrics data helps reinforce their dedication to an improved process safety management system. If the data is collected and no action or review is done, it may send a signal throughout the organization that the metrics program is not a priority. Conversely, if action is taken and priorities are adjusted based on the results of the data, this will reinforce the importance of the program. One way to increase visibility of the metrics program with management is to add a review of all process safety metrics as a standing agenda item during management meetings.

5.6.3   Monitor and Adjust

Sometimes definitions and triggers for data collection can be misinterpreted by the diverse group charged with data collection if training is not sufficient or if the personnel responsible for collecting are not focused upon collecting the data properly. To ensure collected data is accurate and proper, periodic spot-checking data entries from a statistically significant data population will confirm the data quality. A few examples of confirming data integrity include:

- Data entries should be recorded, identifying the participating facilities to confirm that all facilities are submitting data appropriately.

- Design features can ensure consistent data entries when using network databases, such as drop-down menus or predetermined data entry formats (number vs. text).

- Data collection methods at individual facilities should be assessed to ensure that the proper techniques are used. Audit the procedures.

- Alternate sources of data can be cross-checked to ensure that the data is complete. Other company systems may also collect similar data, and cross-checking data entries can verify some of the information.

- Senior process safety specialists with knowledge of facilities can review results and identify implausible results.

- Periodic auditing of data in the metrics reports will confirm the data source. Spot-check the source data with reported information.

- Monitor metrics for unexplained changes and identify a reason for the changes. If the number of MOC reviews drop, but operational activity level remains the same, this may indicate that MOC reviews are not recorded (or not being done).   Alternately, the scope may have crept from the original boundary of the metrics program. If the MI completion rate suddenly increases, this may be due to inspections out of scope being counted as part of the process safety metric, such as the elevator performance metric (not process safety–related) being inadvertently counted as a process safety performance metric.

- Be cautious of unintentional or intentional manipulation of the metrics data. Management leadership should nurture an operational culture that values accurate data reporting, and they should make it clear that intentional manipulation of metrics data is unacceptable. Whether intentional or not, data reported in the metrics system may be compromised by a number of factors. Unintentional errors can be made due to lack of understanding or education about metrics, and this should be addressed in training. Intentional manipulation may arise from a desire to avoid reporting unfavorable results, often called "gaming" the system. For example, continually resetting the completion date for an action item so that it does not show up as overdue will manipulate the metric for tracking the percentage of action items overdue. System management can avoid such manipulation by requiring site management approval and/or an MOC to be completed prior to extending each action item completion target date. Conversely, there may be legitimate reasons to modify data collection requirements, such as excluding overdue training for employees on long-term disability. Communicating and training a well-defined scope for metrics, in addition to validation of data, will help overcome metric gaming.

- Using third-party review and validation of data may be beneficial.

The system integrity depends upon reliable reporting of data. Incomplete information may cause a building trend to be missed or lead to improper conclusions about where weaknesses exist. Conducting audits of the data collection at various locations increase confidence that data is collected appropriately. Self-audits can be helpful in identifying problems, but some portion of the audits should be performed by parties not involved in the data collection. Information from such audits can show how well the metrics system is working and identify improvement opportunities including data system design changes. Undertaking such activities during periodic process safety audits may provide an opportunity to complete a metrics audit.

When developers of a metrics system begin their design work, many assumptions are made about the value of data collected. However, through subsequent experience, some of the assumptions may be proven incorrect or the value of the data collected is not as expected. Reviewing the data periodically and matching the expected benefit and the benefit actually received from that data presents an opportunity to enhance the value of the metrics system. Upon critical review of the data collected, management may decide to stop collecting some metrics since the expected value of the information has not materialized. This will help moderate the workload for those who are charged with collecting and analyzing the data. This may also provide a means to discontinue a low-value activity and replace that activity with a more productive one.

Continuous review of the results from the metrics system should occur to ensure that expected results are materializing. If improvement is not seen over a reasonable period of time, ask if the correct metrics are being collected. Realizing that the metrics' definitions may not be yielding expected value creates an opportunity to improve the metrics system.

5.7 REEVALUATE METRICS BASED UPON EXPERIENCE

The well-known improvement cycle, "plan-do-check-act," embodies the principle of periodically assessing the improvement system progress toward the objectives. It represents a continuous improvement cycle that holds true for metrics. Once metrics are selected, a periodic evaluation should be completed to determine if proper and sufficient information is captured as expected. The original plan may call for tracking a specific characteristic of a process safety element with the expectation that collected information will provide a clear path to improving performance. In the course of collecting data, the initially selected characteristic may not meet expectations for needed information or it may be more appropriate to select another or an additional characteristic. For example, during system implementation a metric associated with personnel hours dedicated to implementation activities may demonstrate whether sufficient resources are being applied to implementation as intended. However, after major implementation progress is made, approaching full implementation, further collection of personnel hours dedicated to implementation may no longer have value. To gain the most from a metrics program, re-evaluating the metrics selection periodically using the improvement effort experience will strengthen the improvement program's efficiency and efficacy.

5.8 CONCLUSION

Implementing a metrics program requires strong management support and leadership as well as a detailed implementation plan. Success will require commitment by those who make business and resource allocation decisions for the organization. Management needs to be engaged regularly throughout implementation to reinforce the reasons for establishing the metrics system as well as to monitor the progress. Unless management support is sustained throughout implementation and subsequent operations, implementation will suffer, if not fail entirely, as resources are shifted to competing activities or other priorities are stressed.

REFERENCES

U.S. Department of Energy, "Training Program Handbook: A Systematic Approach to Training," DOE-HDBK-1078-94, Washington, DC, August 1994

5.7 REEVALUATE METRICS BASED UPON EXPERIENCE

5.8 CONCLUSION

REFERENCES

6

COMMUNICATING RESULTS

Communicating with those who can and do influence process safety is a key to improvement. Many individuals across an organization share objectives to improve process safety. Metrics are a valuable resource to everyone involved in improving performance, and communicating results is a key activity in maintaining a successful improvement effort. Effectively sharing results in a timely manner is the central objective for the communications strategy, and it is important to plan how results will be clearly communicated. Communication strategies should consider the needs, objectives, roles, and responsibilities of the various audiences.

> *Communicating process safety results is a critical element for a process safety improvement strategy.*

Answering a variety of questions that focus on why, who, what, when, and how will help in designing communications. For example:

- What is the value of the metric for a given audience? What would you expect them to do as a result of receiving the information?
- What is the correct amount of data for a given audience?
- Should the metrics be aggregated or detailed? Is data presented as a percentage or normalized value? As an absolute value or as a delta compared to previous reports?
- What is the proper frequency of communication for an audience?
- What is the most effective communication strategy for each audience? Will detailed hard-copy reports or e-mail summaries or some other means be used?
- What is the proper balance of effort and value for providing each report?

As discussed in Chapter 7, metrics should be tailored for the intended or required improvements. The metrics communication plan should be designed to deliver appropriate information to intended audiences so they can make good decisions. This chapter discusses the needs and techniques for communicating process safety results across the organization. Topics include selecting the appropriate metrics to share and techniques for establishing the communications strategy.

6.1   COMMUNICATION ANALYSIS

There are numerous communication models, ranging from complicated and resource-intensive to simple activities using minimal resources. To summarize communications in the most simplistic sense, information is routed from the sender to the receiver. How much information is sent and how the information is received and perceived are only some of the aspects to consider when analyzing communication of process safety metrics.

6.1.1   Identify the Target Audience for Each Metric

When developing the implementation strategy, it is important to understand which metrics need to be reported to which individuals, since not everyone in an organization will use or have a need for every metric. There are a number of internal audiences that have an interest in process safety system performance, and different audiences will use different information to perform their function. For example:

> *Developing Metrics Reports:*
>
> * *How do metrics relate to performance goals?*
>
> * *What data will be reported as "raw" data?*
>
> * *What data is aggregated or normalized, and if so, how?*

- The production manager will likely be interested in knowing the process safety near misses for his department, whereas the shipping manager may not be interested.

- The process safety and maintenance departments may have interest in the quality of completed hot work permits, whereas the business leader may not have use for that information.

Chapter 7 describes the needs of different audiences. An analysis of who needs what should be developed to guide the communications strategy. During early phases of implementation, communications should be analyzed considering various aspects as discussed in the next section. The different information needs of different audiences within the organization are graphically depicted in Figure 6.1.

Figure 6.1 Process Safety Metrics Hierarchy (BP, 2009)

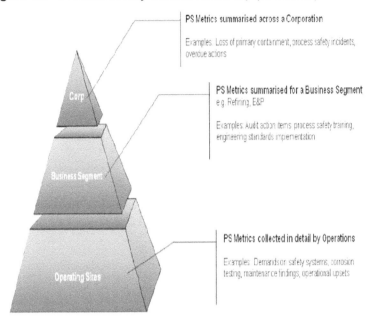

PS Metrics summarised across a Corporation

Examples: Loss of primary containment, process safety incidents, overdue actions

PS Metrics summarised for a Business Segment
e.g. Refining, E&P

Examples: Audit action items, process safety training, engineering standards implementation

PS Metrics collected in detail by Operations

Examples: Demands on safety systems, corrosion testing, maintenance findings, operational upsets

Figure 6.2 also illustrates the way metrics developed for one tier of the organization may support metrics for another tier. There may be some process safety (PS) metrics that roll up through the organization, while others are unique to the performance evaluation needs of that level.

6.1.2   Define Interpretations for the Metrics Data

As the metrics data is collected, defined uniform methods to interpret what the data mean should be developed. Allowing ad hoc data interpretations may lead to divergent and conflicting conclusions, and could result in unintended and inconsistent decisions. Thought should be given as to what the data will indicate, and those interpretations should be provided to data recipients. Of course, experience may indicate that some initial interpretations are not entirely accurate, and the metrics system should have provisions to address and correct such issues.

Figure 6.2 Process Safety Metrics Reporting
(Unpublished BP Presentation)

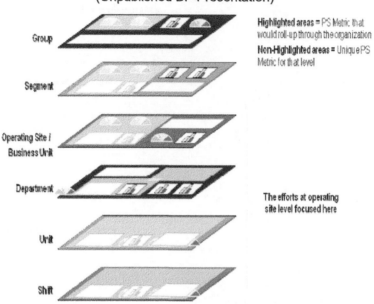

6.1.3   Reporting Results

Reports should be issued in a timely fashion and be consistent with the needs of the audience. Design the report timing and frequency to meet the needs of the audience for each metric.

6.1.3.1   Reporting Frequency

Often the timing of a report is dictated by recipients, and the frequency for many reports is dictated by those who need the data to fulfill their responsibilities. Determining factors include the needs of managers, supervisors, and upper management, and regulatory requirements and requirements of external organizations to which the company belongs. However, the report timing for all purposes should be based on the value of providing that data at the proper frequency. Weekly reports may be appropriate for some purposes, while other reports may only be needed quarterly. The system designer and the intended audiences should jointly decide on reporting frequency. There is little value in reporting metrics more frequently than an audience finds useful, and timing should be suitable for the intended purpose. For example, do not report data daily if the objective is to identify trends over several months. Issuing reports more frequently than useful is distracting and may result in losing the attention of the intended audience. Conversely, not reporting information frequently enough may leave the organization vulnerable to slow recognition of an adverse trend and delay corrective action.

6.1.3.2 Communications Timing Analysis

An analysis of communications timing for each metric should be developed during system implementation planning. A simple matrix tool can be used to document the initial plan for reporting metrics. The communications target matrix can be expanded to include reports' timing, as illustrated in Table 6.1. Experience over time will likely lead to changes in reports' timing as recipients learn how to use the data. The metrics system should include provisions for modifying the initially scheduled communications.

Table 6.1 Communications Target / Timing Matrix Example

	# Process Safety Incidents	# Process Safety Near Misses	% Overdue Personnel Safety Training	Individual Is Overdue for Safety Training	# Safety System Activations	# Overdue Inspections & Testing
Business Executive	√					
Report	Qtly					
Site Manager	√	√	√			
Report	Wkly	Mthly	Qtly			
Unit Manager	√	√	√	√	√	√
Report	Wkly	Wkly	Mthly	Mthly	Mthly	Wkly
Process Safety Professional	√	√	√		√	
Report	Mthly	Mthly	Qtly		Mthly	
Shift Foreman	√	√	√	√	√	
Report	Wkly	Wkly	Wkly	Wkly	Wkly	
Maintenance Supervisor	√		√	√	√	√
Report	Wkly		Wkly	Mthly	Mthly	Wkly

6.2   SELECT APPROPRIATE COMMUNICATION CHARACTERISTICS

As the communication strategy for process safety metrics is formulated, consider various factors to ensure that the communications are appropriate and effective. Such characteristics include:

- Select appropriate metrics for each audience;
- Select the appropriate number of metrics to communicate;
- Determine the frequency of metrics communications;
- Decide when normalized metrics are used;
- Consider organizational culture;
- Balance value and effort for providing metrics reports; and
- Periodically evaluate the value of communications.

Fashioning a well-thought-out communications strategy will help ensure effective motivation for personnel in supporting the process safety improvement efforts. Choosing the communication characteristics for each audience is important to successfully communicate the proper messages.

6.2.1   Select the Appropriate Metrics for Each Audience

Various audiences will use different metrics as introduced in Section 6.1.1. Communicating the right metrics to the intended audience will help the recipients meet their responsibilities. If the metrics reported to a specific audience are not relevant or do not provide the desired information, then the communications will not be valued. For example, if detailed metrics are reported to an audience expecting high-level metrics, they will probably not review the detailed data nor understand trends within the data. Data that is aggregated may not provide sufficient information for an audience needing detailed data.

6.2.2   Select the Appropriate Number of Metrics to Communicate

Providing the right amount of information to an intended audience facilitates information transfer and makes the communication more effective. Reports that are easy to review and understand and provide needed information are appreciated by recipients and allow them to efficiently assimilate the information.

Overloading audiences with too much information may lead to confusion and seem overwhelming to those who do not have the knowledge to fully understand all the data. Providing too much data may require a recipient to sort through a lot of data that might not be relevant to his or her responsibilities. If this takes a significant amount of time, it is likely that the recipient will not bother to do this routinely.

Processing large amounts of data will be time-consuming for the personnel analyzing data and formulating the metrics reports, and processing too much data can lead to inefficient utilization of resources. Selecting the optimum number of metrics to be processed and reported to the intended audiences will facilitate the communication process.

6.2.3 Determine the Appropriate Frequency of Communication

Sharing metrics information on an appropriate frequency is important for successful communications. Communicating metrics too frequently or not frequently enough can be ineffective and possibly detrimental. Those audiences that have a need to understand longer-term trends will neither need, nor appreciate, updates of information that are too frequent for their needs. Sometimes such reports are placed aside and not reviewed. Important information that is interspersed among a large number of reports may be overlooked even though the information deserves special attention. On the other hand, those who are responsible for day-to-day operations will use frequent informational updates, particularly those recipients who will make immediate changes based upon the information. Personnel that deal with day-to-day operations cannot wait weeks or even days for some information. Making a special effort to keep the communications fresh and appropriately timed for each audience will facilitate reliable information transfer.

6.2.4 Decide When Normalized Metrics Are Appropriate

Normalized performance metrics provide a means to compare performance across similar as well as diverse groups, as described in Section 3.3.2. Normalized data provides a basis for communicating performance gains or deficiencies for diverse operations. These metrics can provide a broad-based view of performance and allow straightforward and simple communication messages relating to all personnel.

6.2.5 Consider Organizational Culture

Operations of multinational organizations often span multiple countries and cultures, and their broadly distributed metrics reports must be appropriate for all intended audiences. Using specific language in one culture may have unintended and problematic connotations in another, so the metrics report designs must account for cultural sensibilities. Metrics training across diverse cultures may be needed to ensure understanding of the intention and content of metrics reports.

Some companies have extensive experience in performance improvement programs, and their normal communication style is to report the number of defects or failures. This communication style would routinely report the number of process safety system failures by element. Others may consider using a more positive message where the focus is on percentage of successes, such as a reporting a 99.5 percent success rate for all process safety system activities. However, using this style may not provide enough detail since a large facility with hundreds of activities occurring every day, even 0.5 percent failure rate represents a large number of failures. In more hierarchical organizations, communications may be very direct and focus only upon failures and corrective measures, stressing personal accountability. Other companies report data in a neutral fashion, focusing upon system weakness that permitted a failure. Whatever style the organization adopts, it must recognize potential shortcomings of communications and ensure that gaps will not develop.

In any case, if the organization's culture is deemed to be tolerant of incidents and near misses, communications may need to focus on encouraging a more disciplined culture that does not accept such deviations. The delivery of the metrics message must properly characterize performance and inspire personnel to become part of the process safety efforts. Even the best metrics will not compensate for poor corporate safety culture. Changes in culture must be driven from the top throughout the organization. Nevertheless, good metrics will provide the data management can use to evaluate and track expected behavior.

In an organization that has not historically evaluated performance using objective data, the initial reported results may appear to indicate that performance has declined. This sometimes occurs when the historical perspective is too optimistic. For example, the number of near misses often increases as reliability of reporting improves. The recipients for metrics reports should be advised what to expect and provided explanation as to why the results may not meet expectations. The potential cultural impact, if left unaddressed, may raise doubt about the accuracy and value of the metrics reports, and these potential negative reactions need to be managed.

6.2.6   Balance Value and Effort for Providing Metrics Reports

As the metrics reporting is designed for audiences, the amount of effort and resources needed to generate the metrics reports should be balanced with the value derived from the reports. Management at various levels, from site managers to corporate management, will dictate many reports. As objectives for these reports are agreed upon, the effort required to generate the reports should be estimated and approved by those requesting the information. Some audiences will need aggregated data reports, and this activity will require analysis and preparation of the raw data. Resource estimates to generate these reports should be included in the funding requests for the metrics system. Often, an unbiased analysis of data requests can improve the request, requiring fewer resources and still providing needed information. Automation of data reports, using computerized databases, can often save ongoing staffing requirements after an initial investment in the software.

6.2.7   Periodically Evaluate the Value of Communications

As time progresses and process safety system performance improves, some data may no longer be useful. Periodically evaluate the data that is captured, analyzed, and reported, and determine if the data is still valuable to the improvement effort or to maintaining performance. Remove data that has proved to be unimportant or has outlived its usefulness. Add new data that can improve the ongoing evaluation of performance. As the process safety system performance improves, it follows that the appropriate metrics will change, and it is important to keep the metrics relevant. Continuing to report information that no longer reflects the current status of process safety performance, or no longer distinguishes changes in performance, may be misleading and may discourage recipients from using the reported metrics information.

6.3 REPORT APPROPRIATE DATA TO DIFFERENT AUDIENCES

Communicating process safety metrics should match the needs of the intended audiences to maximize the benefit for them, and the data communicated will vary depending upon those needs. Metrics reports for upper management usually contain aggregated or normalized data, focus on trends, and are issued on a periodical basis. Personnel that use the results to carry out their day-to-day responsibilities usually need detailed data that is reported frequently. The data reporting frequency is usually dictated by the audience needs, as discussed in Section 6.2.3, and timely reporting is important for effective communications.

The metrics system designers need to define the metrics communication methods to be used for each audience. Some audiences will benefit from written summary reports that contain recommendations for change based upon data analysis. Others may use raw data collected from automated data collection systems or reports that are generated using algorithms that show specific trends. The example communications matrix in Table 6.1 can be modified to include not only the information conveyed but also the communication method to be used. Agreement between the report designer and each audience will be critical for a successful communications strategy.

6.3.1 Metrics Reporting for Operations, Craft, and Supervision

Operators, craft, and maintenance personnel, and their supervisors, use metrics to monitor the operating process, the maintenance and reliability systems, and process safety performance. Their needs are frequently short-term information that identifies when prompt attention is needed for a given situation within their area. The data for this audience is often detailed in nature and relates to specific operating or maintenance work tasks. For example, all training requirements are usually listed by specific job function, and completed training is tracked for each individual using a training database. Safety training data could be extracted and exported to the process safety metrics system. Individuals will have interest in how their work group and area are performing. Reported metrics will likely include percentage of training completed on time and other information deemed appropriate. Reporting process safety metrics to individuals provides an indication of their performance as well as reinforces improving and maintaining process safety. Span of control will also influence information contained in metrics reports. For instance, a supervisor will have interest in training completion metrics for the site, but that same supervisor will be most interested in the performance of his shift or crew. The supervisor's focus is likely percentage of training completed by the group and the names of individuals who are delinquent in training.

Since the amount of potential data may be quite large, the designer of the communications reports and an intended audience member should work together to identify what data is relevant for process safety performance monitoring, and map out the data sources for that information as described above in Table 6.1. Such mapping could be done with the mechanical integrity coordinator in an operating unit, and provide answers to the following:

- Identify the data that provides the desired performance measure
- Identify where the data resides
- Include that data in the unit's process safety performance report

Computerized data management systems have the capability to automatically construct data reports for process safety data analysis once the report parameters are defined.

6.3.2   Metrics Reporting for Line Management

Line managers use a variety of metrics to monitor their organization's performance and progress toward meeting the objectives for which they are accountable. Their data reports should include appropriate indication of the organization's process safety performance and directly relate to the line manager's objectives for the organizational unit. Some call this the "scorecard" for the organizational unit, and it will likely include operating and maintenance metrics as well as process safety performance indicators. Examples of process safety performance indicators may include but are not limited to:

- The number of process safety incidents;
- The number of process safety near misses;
- The number of overdue PHAs;
- The number of unresolved PHA and MOC recommendations;
- The number of overdue and/or missed inspections;
- The number of individuals with past due training;
- The number of outstanding process safety incident updates;
- The number of challenges to automated safety devices, such as safety instrumented systems; and
- The number of excursions outside of defined safe operating limits.

Aggregated metrics can be useful in tracking trends in performance over time and may substitute for continuously monitoring the detailed operations. However, the line manager will likely need to track some detailed data that indicates performance of special interest, such as data related to correcting a known performance deficiency. The line manger will use a variety of data, and the report designer must understand the line manager's needs and provide the appropriate data.

6.3.3   Metrics Reporting for Process Safety Professionals

The process safety professional will have primary interest in the data that indicates process safety system performance. This audience includes both process safety professionals at a given site as well as those who have multiple-site or corporate responsibilities. Process safety professionals often document and report selected metrics, but much of the raw data is generated by others, such as operations and maintenance personnel. It is important for the process safety professional to reach agreement with those generating the data on how that data is transmitted to him or her. Issues to resolve in developing a data exchange plan include, but are not limited to:

- Who will collect, validate, and report the data?
- Will automated data collection and reporting be used?
- What data or information will be reported?
- What is the reporting format for the data?
- What is the frequency of specific data reports?

Computerized databases can facilitate data collection as well as generate and distribute reports. Operating system information, such as the number of challenges to process safety interlocks, can be automated using digital process control systems; the process safety professional can have access to this information for monitoring operations performance from a process safety perspective. If the digital process control system has networking capability, reports can be electronically sent to the process safety professional automatically.

Process safety professionals in corporate organizations will use detailed information to track performance of the process safety systems across the enterprise. Often, database reports from parallel organizations, such as data from safety, health, and environmental departments, can be used to confirm the validity of some process safety data collected. For instance, environmental reports usually document loss of containment incidents that may not have been properly classified as a process safety incident. Similarly, the occupational injury and illness reports may capture injuries related to the operating process that were not reported as process safety incidents. The process safety professional should recognize opportunities to receive data from parts of the organization that are not directly responsible for process safety performance but generate information that might assist in assessing the overall process safety performance. For example, the human resources department may already collect data on staff responsibilities, training resources and records, and staff turnover.

6.3.4   Metrics Reporting for Facility Management

Facility managers use a variety of metrics to monitor operating, maintenance, and process safety performance at their site. These metrics often include a consolidated or aggregated view of metrics as well as some selected detailed data of interest. These reports can indicate the process safety system performance across the site and provide a performance comparison for individual operating units. This allows the site manager to compare performance among the site operating units and use this information to focus performance improvement efforts. Metrics reports often include but are not limited to items such as:

- The number of process safety incidents;
- The number of process safety near misses;
- The percentage of personnel with past due training;
- The percentage of past-due equipment and instrumentation inspection and testing; and
- The percentage of past-due PHAs.

As mentioned earlier, metrics reports can be viewed as the scorecard for the site and provide a means for the site manager to measure progress on meeting the objectives and results for which he or she is responsible.

Reports for site management should be frequent enough to provide information for timely decisions on improving performance and correcting identified deficiencies. The site manager is often more interested in trends and sustainable performance level indicators than in short-term variations that may obscure the long-term performance level. The depth and frequency for reporting selected metrics will vary depending upon the priority issues facing the site manager. Reporting methods for site managers can vary from hard-copy reports to automatically generated reports from computerized databases. The metrics report designer and site manager need to develop the objectives and framework for the communication specifics.

6.3.5   Metrics Reporting for Corporate/Business Management

The primary interest of corporate and business leadership is in meeting broad-based process safety goals as described in Chapter 5. They need to understand if the operations within their areas of responsibility are meeting expectations and agreed-upon improvement objectives. Corporate and business leaders not only track aggregated performance that transcends individual sites, but also use selected metrics to make performance comparisons among sites under their oversight.

Perhaps most important, corporate leaders should receive information about areas where there are significant performance gaps—sites, businesses, etc. that are performing poorly, yet such information might be lost in averaging of aggregated data. Management reporting systems must highlight the outliers. For example, a plant may have an acceptable average overtime rate without appreciating that the rate is driven by a few individuals who work an extreme number of overtime hours.

Metrics reports used by corporate and business leaders contain less detailed information than used by site and line managers, and these metrics reports often share the following common characteristics:

- Aggregated or normalized metrics
- Results linked to performance objectives
- Focus on trends
- Reports issued on a periodical basis, such as monthly, quarterly, or annually

High-level leaders sometimes track specific, detailed metrics to understand progress for specific objectives that deal with high-priority issues. The report designer for this audience must consider a variety of issues, and often differently structured reports are needed for individual leaders to suit their needs. Issues to address for the specific reports include:

- What data is aggregated or normalized? If so, how?
- How do metrics relate to performance objectives?
- What data will be reported as "raw" data?
- What is the granularity of data reported, such as reporting by business unit, by facility, or on some other basis?
- Is data presented only for a given time period or is it presented within a historic perspective illustrating trends?

Upper management will need to be coached about the metrics and how to react to the metrics reports. One organization should be praised for bringing their metric from 65 percent to 85 percent while another should be questioned about why a metric slipped from 95 percent to 90 percent. This requires management to be knowledgeable or to be coached before making comments about metric performance.

6.3.6 Metrics Reporting for Board of Directors

Communicating process safety performance to the board of directors is usually in the form of normalized metrics that indicate long-term trends. The board is accountable for long-term performance and risk management of the enterprise. An annual report is often made to the board, with more frequent metrics relating to high-priority concerns spanning the enterprise reported throughout the year or as necessary to provide the board with the information needed for decision making. In addition to reporting *to* the board, the organization may also report on metrics and actions taken *by* the board that demonstrate delivery on obligations.

6.3.7 Metrics Reporting for External Stakeholders

Some companies and industry associations publicly share their performance metrics, and it is common for companies to report results for their environmental, health, and safety programs. However, few companies report process safety results, and adding process safety results to annual reports is a desirable step forward in transparency for performance. Such reports are valuable in developing good relations with the public, advocacy organizations, and regulatory agencies. However, selection of metrics and their presentation must be considered thoughtfully. When reporting company performance data in an aggregated format, the report designer should consider the following:

- What information is appropriate for external reporting? Who will decide what is to be reported publicly?

- What granularity is used for the reported data? Will reports only include performance for the company overall or by business or by site?

- Is data presented only for a given year or is it presented with a historic perspective?

- Is the data normalized? If so, how?

Many companies participate in industry organizations that require reporting of specific process safety metrics. These industry organizations define criteria and collect and publish process safety data. For example, the member requirements for American Chemistry Council Responsible Care® require reporting process safety incidents (ACC, 2009). The company needs to decide if it will publish information beyond that required by an industrial organization.

6.4 TOOLS FOR COMMUNICATING METRICS

Many tools exist to aid in disseminating process safety system metrics among various audiences. These tools range from paper communiqués distributed widely across an organization to modern electronic tools designed to reach many persons simultaneously. Companies often use a variety of methods to communicate and to reinforce important learnings from the metrics. Communication methods are often formulated specifically for target audiences where there is a perceived advantage of one method over another.

6.4.1 Establish Networks to Share Results

Many organizations foster communication networks to facilitate the delivery of the organization's objectives and track progress. These same networks can be useful for communicating information about process safety metrics and providing a broad-based resource to judge the improvement efforts. Such networks can work especially well with process safety professionals and operating personnel involved in process safety. As practitioners share their experience and the results from their improvement efforts, the transfer of information can be inspiring, as improvements are made real. The continuous communication of metrics data to the network members provides information for advocating the improvement efforts in which they participate.

Communications within a network are usually two-way, and this is an advantage since a dialogue among interested parties on specific issues occurs. Questions arise during such communications, such as:

- Why was the effort undertaken? What did you expect to gain?
- How did performance improve? Is it sustainable?
- What technique(s) was used?
- How did you start? What was needed to be successful?
- When will you finish?

These dialogues lead to sharing useful techniques and ideas. Many networks are informal in nature, being developed by parties with common interests. However, it is not unusual for some networks to be formalized by owners of the improvement effort.

6.4.2   Discuss Safety Performance Routinely at Meetings

Some companies start every meeting with a discussion of safety performance. This originally began with occupational safety but can be extended to process safety being discussed on some frequency. Often the focus is on the metrics collected and the organization's performance as compared to objectives. Continuously repeating such information and the actions being taken to improve performance reinforces the organization's commitment to safely operate its processes. It is powerful when the CEO and other high-level managers begin meetings by briefly reviewing the organization's safety performance. This technique has been used routinely in many operating sites as construction workers began each shift with a "tool box" meeting to review important safety information before beginning work. Operating personnel have also used a similar technique at shift change to review timely issues for the oncoming shift. Integrating process safety performance and metrics into such meetings can aid in reinforcing the importance of process safety at all levels of an organization.

6.4.3   Mass E-mail Distributions

Some companies use mass mailings via e-mail to reach most of their personnel, and sending out important information via e-mail is becoming more prevalent. Some utilize a simple format to transfer information learned during investigations for incidents and near misses, as collected lagging metrics indicate a significant issue for many to address. Using e-mail is a rapid means to reach a large number of employees, but judicious use of this technique should be exercised since a high number of e-mails may lead to complacency and consideration of the e-mails as "spam." Some external information can also provide a subject matter for e-mails on metrics and performance trends. For example, the American Chemistry Council

(ACC) and the American Petroleum Institute (API) publish reports that consolidate reported metrics from their organizations' members, and these summaries can provide useful comparisons to an organization's own experience.

6.4.4 Electronic Information Systems

Reporting data can be automated using computerized databases. These database systems can generate aggregated data reports and provide predetermined graphical presentations for trend analysis. Often, the reports for corporate and business management are vetted by process safety professionals prior to publication to ensure that there are no data-processing problems and that the trends are presented properly. Some companies use computerized databases to monitor process safety incidents and unresolved audit findings; these systems are normally capable of generating metrics reports. These systems not only can collect metrics data and formulate database reports but can also automatically communicate the data to selected recipients.

6.4.4.1 Digital Process Control Systems

Digital process control systems not only control the operating process, but can also collect specific data and trend information. These systems can be programmed to recognize aberrations in the process operation and record the information. Some systems have the capability to automatically structure reports and transmit those reports to recipients via an organization's information network. Metrics related to the operating process can be automatically collected by the digital process control system and transferred into a database for accumulation and processing.

6.4.4.2 Process Safety Databases

Many process safety information databases are designed to share metrics across a facility or even an entire enterprise. For example, databases developed to monitor PHA studies across an enterprise can track PHA studies by business, by site, and by unit. Systems are developed by the operating company or purchased from a systems vendor, and some enterprise business software can provide similar capabilities.

A strength of these databases is the automated reporting capability. Most databases can generate reports that provide an organized view of data and can provide trending information. These are powerful tools for the process safety professional who will be reviewing a lot of data and wanting to understand the strength and weaknesses in the process safety system. It is important to understand the investment, time, and resources committed to a database system since many systems are costly to implement. Deciding the needs for data management capabilities will depend upon the scope of the metrics systems and factors such as:

- How many facilities are included in the metrics system;
- How broadly are data contributors dispersed (geographically); and
- How much benefit is derived from automatic data reporting by reducing human resource needs.

The metrics of interest for the process safety professional will cover many, if not all, process safety elements. The inputs may come from a number of units across a plant site or from multiple plant sites around the world. Design and development of the reporting system needs to take into account all these factors.

6.4.4.3 Dashboards

Some computerized systems allow constant viewing of selected data across digital networks, and this functionality is often referred to as a dashboard. Placing data into a well-designed presentation or dashboard allows the user to view the latest data at any given time. Often the user has the capability to select which data to display based upon his or her specific interest. The time lag between actual activity and when the dashboard displays information is a function of the system design, relating to how the data is collected and is available in the database. Some dashboards contain real-time information generated by a process control system and can be made available across a digital network. Dashboards can be useful for supervisors and managers who may not continuously monitor the detailed process operations but are interested in following specific parameters. Additionally, dashboards can provide data trending over selected time periods. One dashboard feature that many find helpful is a visual indication if the data is within expected limits, such as "green light," or using "yellow light" or "red light" for parameters that are outside of predetermined limits.

Some companies have a suite of metrics that they update weekly, monthly, or quarterly depending upon their needs. This data is usually system performance metrics that summarize the status of activities or metrics determined to be of interest. These systems can provide process safety information as soon as it is developed and entered into a network database. With the proper access, operators and craft personnel and their supervisors can view this information, and nearly any audience can view the reported data in a predetermined format.

Dashboards can be designed using simple tabular data presentations or using a sophisticated network-based system that allows the individual to select what data they wish to display. Design decisions are a function of the organizational needs and the available resources.

Information systems with less data will tend to depend more upon standard tabular formats using manually manipulated data; an example is shown in Figure 6.3 on the next page. Large organizations that want data reporting over multiple units or sites will tend to invest in more sophisticated systems. Example dashboard systems can be found on the accompanying CD.

6.5   CONCLUSION

The best metrics are useless unless they are communicated in a manner that is useful to the recipient in terms of frequency, form, and content. Therefore, metric communication strategies must identify the audiences for the metrics and their specific needs, objectives, roles, and responsibilities. Common audiences for metrics include business executives, site managers, unit managers, process safety professionals, shift foremen, maintenance supervisors, and, of course, the public.

Reports to each audience should be issued in a timely fashion and be consistent with the needs of the audience with respect to communications frequency, number of metrics to communicate, and the form of the metrics (actual, normalized, or both). Such decisions are driven by the responsibility of the audience within the organization and organizational culture.

A variety of communications tools may be used to disseminate metrics and metrics reports. Such tools range from paper communiqués distributed widely across an organization to modern electronic tools designed to reach many persons simultaneously. Common communications vehicles include establishing internal networks to share and discuss results, discussing safety performance routinely at meetings, distributing mass e-mail within and even outside the company, and creating electronic information systems such as dashboards to make information broadly available.

The metrics communications strategy—audiences, content, and dissemination form—should be periodically evaluated to assure that it continues to meet the needs of the audiences and the organization.

Figure 6.3 Example Dashboard Using Tabular Format

		PSM						MOC		Standard Operating Procedures		PSM		
		Number A/B/C incidents for the month just ended	Number PSM A/B/C incidents YTD	Number closed PSM A/B/incidents YTD	Number Level D PSM incidents and misses for the month just ended	Number Level D PSM incidents and near misses YTD	Number currently past due, PSM-related incident recommendations/corrective actions	Number PSM MOCs open >90 days past actual start-up date of the change	Number PSM MOCs initiated YTD	Number of PSM SOPs that have come due for their recertification preceding 12 months	Number PSM SOPs overdue for recertification during preceding 12 calendar months	Percent of employees who are overdue in receiving refresher training on PSM SOPs	Lowest functionally trained team %	Average % of teams functionally trained for complete operational coverage
		May	May	May	May	May	May	May	May	May	May	May	May	May
A	Chem Ops	0	0	0	2	11	0	0	40	5	0	0	188	241
	Finishing	0	0	0	0	0	0	0	22	6	0	0	433	458
B	Latex	0	0	0	0	0	0	0	4	6	0	0	100	150
	Resin	0	0	0	0	1	0	0	27	122	0	0	100	150
	Tank Farm	0	0	0	0	2	0	0	11	15	0	0	200	200
	Finishing	0	0	0	1	9	0	0	21	0	0	0	125	214
C	Chem Ops	1	1	0	0	2	0	0	5	19	0	0	200	200
	Finishing	0	0	0	0	1	0	0	10	11	0	0	150	150
	PBD	0	0	0	0	0	0	0	0	5	0	0	200	200
D	Finishing	0	0	0	0	1	0	0	3	0	0	0	200	535
	Latex	0	0	0	0	1	0	0	11	8	0	0	113	259
	Resin	0	0	1	0	5	0	0	12	19	0	0	210	243
	Tank Farm	0	0	0	0	1	0	0	4	7	0	0	271	271
E	Compounding	0	0	0	0	3	0	0	12	17	0	0	314	356
	Resin	0	0	0	0	1	0	0	24	24	0	0	241	311

REFERENCES

American Chemistry Council, "Responsible Care®: Process Safety," *n.d.,* available at *http://www.americanchemistry.com/s_responsiblecare/sec_members.asp?CID =1318&DID=4861*

Broadribb, M. P. et al., "Cheddar or Swiss? How Strong are your Barriers? (One Company's Experience with Process Safety Metrics)," Presentation at CCPS 5th Global Congress on Process Safety, Tampa, Florida, April 26–30, 2009

7

USING METRICS TO DRIVE PERFORMANCE IMPROVEMENTS

The identification and implementation of new or improved process safety metrics alone will not improve process safety. Rather, the data must be collected, analyzed, communicated, understood, and *acted upon*. While metrics can highlight process safety performance strengths and weaknesses, process safety improvement comes when such efforts are directed towards achieving and maintaining desired performance by correcting weak performance.

Certain metrics will identify weaknesses in performance and even alert personnel to issues that need immediate attention. Managers and supervisors can only take action to correct problems if they are aware of the situation. Management has a major role in reinforcing expectations and holding themselves and other personnel accountable for meeting their assigned tasks and responsibilities.

Management's visible responses to problems identified by the metrics are as important—if not more important—than the metrics themselves. Objectives may be defined and responsibilities assigned, yet a lack of management follow-up can undermine focus and weaken commitments. Activities that

> *Management's visible responses to problems identified by the metrics are as important — if not more important — than the metrics themselves.*

do not merit management review are perceived as unimportant and of little value. Expecting others to meet objectives that are perceived as unimportant to management is unrealistic.

Previous chapters have addressed ways to identify, implement, and communicate new or improved process safety metrics. This chapter discusses strategies and tactics to convert information derived from metrics into actions that improve performance. While specific activities and programs may vary among organization, basic strategies and approaches include:

- Identify weaknesses and performance deficiencies
- Hold responsible parties accountable
- Conduct periodic management review of the system
- Cultivate a positive safety culture
- Commit to managing the safety culture
- Communicate to reinforce objectives

7.1   IDENTIFY WEAKNESSES AND DEFICIENCIES IN PROCESS SAFETY PERFORMANCE

Metrics analysis can indicate potential trends in performance and identify weaknesses in a system or the implementation and execution of the system. Many companies and organizations track incidents within facilities and across organizations. Increased frequency of incidents may indicate multiple weaknesses in the design and/or implementation of the operating and process safety systems. Metrics that track items such as the frequency of procedures not followed may indicate improper training or leadership in the conduct of operations. Metrics that indicate scheduled inspections are routinely late may highlight the need to improve design of inspection schedules or commitment/leadership to execute an inspections program reliably. Metrics that indicate the failure of control instrumentation may indicate the need to improve the instrumentation testing and calibration systems. Metrics that indicate routine delinquency in closing actions from MOC reviews can warn of deficiencies in the MOC system. PHA studies that are chronically overdue may indicate that the planning, resource commitment, and execution of the PHA program need improvement. These are only a few examples of metrics that can be tracked to indicate the performance of the process safety systems.

Once the metrics indicate a problem, managers responsible for the problem area should begin an incident investigation to identify the root causes of the problem and appropriate corrective actions. To assure corrective actions are implemented, managers should establish improvement goals and an action plan—including appropriate metrics—to correct the problem. Using metrics to indicate problems is not enough. The process safety system must also include requirements that actions will be taken to correct problems identified.

Chapter 6 describes formal or informal communication networks that many organizations support. In addition to delivering the organization's objectives and tracking improvement progress, these networks are useful for identifying differences in performance as well as providing a reservoir of practitioners and peers to help analyze causes and strategize solutions. Networks composed of peers, such as process safety professionals or managers from different facilities within the organization, may also foster an incentive to perform as well as if not better than peers as demonstrated by the comparable metrics

7.2 LEADERSHIP COMMITMENT TO PROCESS SAFETY PERFORMANCE

A demonstrated commitment to process safety by an organization's senior leadership is essential. The organization's leaders must set the process safety "tone" at the top of the organization and establish appropriate expectations regarding process safety performance. Those expectations must reflect an unwavering commitment to process safety and infuse into the organization's work force the mind-set that process accidents are not acceptable.

In 1996, Jeff Lipton, CEO of NOVA Chemicals, set a goal for NOVA to reduce the number of uncontrolled process fires to zero as part of an effort to reduce the risk of a catastrophic process incident. From that point forward, process fire targets were routinely reviewed and made part of everyone's objectives—including Mr. Lipton's. He spoke consistently and passionately both to employees within NOVA and to his peers in the chemical industry advocating strong executive commitment to process safety. As a result, NOVA has seen a reduction of uncontrolled process fires from 65 in 1998 to 6 in 2007. The uncontrolled process fire target for 2008 is 5 or less (see Appendix III).

> *NOVA Chemical's CEO, Jeff Lipton, spearheaded an effort to reduce the number of uncontrolled process fires to zero. The number of uncontrolled process fires has fallen from 65 in 1998 to 6 in 2007.*

7.3 HOLD RESPONSIBLE PARTIES ACCOUNTABLE

As described in Chapter 4, in an ideal process safety system, the organization's goals, objectives, and metrics should be reflected across the organization and, as appropriate, in the performance expectations and contracts of individuals with responsibilities under the system. A key factor in ensuring that systems perform as intended is to hold responsible personnel accountable for desired results—objective metrics are one way to measure and document results. Responsibilities and accountability vary throughout an organization, from the shop floor to the board of directors, depending upon the desired result and the ability to influence the outcome.

> *Responsibilities and accountability vary throughout an organization, from the shop floor to the board of directors. These different audiences should be held accountable for taking actions warranted by the metrics they receive.*

Just as different metrics are designed for different audiences within the organization, these different audiences should be held accountable for taking actions as warranted by the metrics they receive. For example, operating and craft supervisors may be held accountable for adherence to procedures and completion of training, while line management may be held accountable for completing MOC reviews and addressing incident investigation actions in a timely fashion. Similarly, process safety professionals may be held responsible for scheduling and executing PHAs, while facility managers may be held responsible for actual incidents and near misses. Finally, senior management and the board of directors may be held accountable for establishing and maintaining a positive safety culture within the organization, and assuring adequate resources are made available for process safety.

Clear definition of responsibilities and individual accountability are important. An individual may have the responsibility to develop and collect metrics information but may not be accountable for the results themselves. For example, the objective for process safety improvement is to lower the number of incidents, yet the individual responsible for collecting process safety incidents data may not be the person to be held responsible for an increase in incidents. The individual responsible for collecting data would, however, be responsible for the accuracy of the data.

7.3.1   Link Process Safety Performance to Personal Performance

There is a strong link between employee performance and reinforcement that individuals receive. Reinforcement (positive or negative) comes in the form of comments and recognition for activities and behavior, as well as more tangible demonstrations such as monetary compensation or advancement. Using the breadth of reward and reinforcement techniques to drive improvement can strengthen process safety performance through a clear system of reward and consequences. Many organizations have linked personal safety performance to rewards, compensation, or advancement for many years but may not have extended these policies to process safety.

Linking process safety results with personnel compensation, however, must be done carefully to avoid counterproductive incentives. The compensation systems must not discourage reporting unsafe conditions or other "gaming" of the metrics system. Management needs to understand, and the reward/compensation system should appreciate, that an organization will frequently see an initial inverse response (results get worse before getting better) in the first year or two of a new metric—not necessarily because performance is deteriorating, but rather because reporting accuracy is improving. Similarly, the increase in some metrics may actually be an indication of *increased* process safety. For example, an increase in reporting near misses or other leading indicators should lead to a decrease in the frequency and seriousness of actual process safety incidences (Phimister, 2003).

7.3.1.1   Rewards for Good Performance

Some companies have used employee reward programs to reinforce a commitment to safety. Such programs may reward individual or group performance. Reward programs can include simple trinkets such as t-shirts or mugs, "perks" like reserved parking places, or even monetary compensation. Many times rewards are presented at plant or company functions and documented in company-wide communications to provide public recognition of an individual's or a group's contributions. As an example, the DuPont Safety Awards were launched in 2002 with the purpose of stimulating individual or collective initiatives for safety enhancement and accident prevention in Europe, the Middle East, and Africa. Prizes are awarded for performance improvement, innovative approaches, visible management commitment, sustainable business impact, and cultural evolution.

Some plant sites conduct reward programs for their employees and stress that the contributions impact co-workers and friends by improving safety. These programs are results-based and often are popular with employees. Process safety metrics can be the basis for awards in recognizing the contributions of individuals or groups who meet, or exceed, expected results. Some organizations provide awards for new ideas and suggestions that provide a new path to improvement; using metrics can aid in selecting valuable improvement ideas.

7.3.1.2   Consequences for Poor Performance

As described in Chapter 5, evaluation of those responsible for the process safety system and their performance should be reflected in individual performance objectives or contracts. Poor performance against process safety objectives should be reflected in annual evaluations.

People must be held accountable for results. If action items are not completed in a timely basis, there needs to be an analysis of why and a plan for improvement. If standard operating procedures (SOP) checklists are not used rigorously, there needs to be an analysis of why with appropriate consequences (punishment should be a last resort). There should be reasons why objectives are not met and those barriers need to be removed. Individual sites within an organization may establish specific process safety goals, activities, and behavior on individual performance contracts, which are reflected in the annual performance review process that links to merit pay raises, bonuses, advancement opportunities, and the like. Consequences for poor process safety performance must not be limited to supervisors or facility managers but must be a factor in performance appraisals throughout the organization.

7.3.2   Use Caution in Linking Personnel Rewards and Consequences to Process Safety Performance

Tying employee performance evaluation, including linking compensation or bonuses, with process safety performance should be implemented only after careful consideration and safeguards are in place to prevent perverse outcomes.

Attaching financial consequences to performance can provide an incentive to manage the metric rather than the actual performance. For example, if performance is measured by the number of employees who have completed training on schedule, the number of employees trained could be increased by decreasing the quantity or quality of the training. While the number of employees trained would go up (the metric would trend positively), the value of the training (increasing safety awareness, improving overall facility safety) could easily go down.

Those responsible for aspects of process safety should be held accountable, but it is important that the system be able to evaluate the cause of poor performance rather than automatically linking it to poor performance. Where there is poor performance with respect to such an indicator, the first reaction should be to investigate the reasons. It may be that there are not sufficient resources to carry out the activity effectively. Penalizing poor performance in these circumstances is bound to lead to perverse outcomes (Hopkins, 2009).

7.4   ENGAGE THE PUBLIC

As described in Chapter 8, companies and industry associations are increasingly sharing their performance metrics with the public through corporate performance or sustainability reports that include information on many regulatory and voluntary areas including environmental releases, carbon footprints as well as safety performance.

Some companies not only are publicizing results, but are actually publicizing their performance goals and challenging the public to track performance and hold the company accountable for achieving the goal. Dow has posted on its corporate Web site the goal that "All sites will reduce process safety incidents by 75 percent and the severity rate by 95 percent, based on the 2005 baseline" and a commitment to "collaborate with our communities to set and commit to local goals" (Dow, 2008a). The company posts year-end and quarterly reports on performance against the goals regardless of whether the news is favorable or not.

> *"All sites will reduce process safety incidents by 75 percent and the severity rate by 95 percent, based on the 2005 baseline. . . . The 2015 goal of 14 is a 75% improvement from 2005."*

Making metrics public can be an especially powerful way of maintaining upper management's commitment since the CEO or other senior managers are likely to be called to account by the public if goals are not met or performance declines. BP America maintains a Web site describing the company's commitment to process safety as well as links to reports and other information on the 2005 Texas City explosion (BP Web site).

Organizations are increasingly making their members' performances publicly available both to demonstrate their members' commitment to good performance as well as to improve credibility with the public. The American Chemistry Council, as part of the annually Responsible Care® report, posts process safety incidents by company as well as industry aggregate for specific years (ACC, 2008). The International Council of Chemical Associations (ICCA) collects and provides information on chemical industry performance measures from 53 countries. Though currently ICCA only reports worker safety performance, it is negotiating process safety metrics that can be adopted across the member organizations (ICCA, 2008).

While used more frequently in the environmental arena, governments and advocacy organizations have been successful in driving performance improvement by using Web site and other public disclosure mechanisms to make information on companies available. One of the best-documented examples of public accountability has been U.S. EPA's Toxics Release Inventory (TRI) (U.S. EPA-TRI, 2008). Created by Congress as part of the Emergency Planning and Community Right-to-Know Act (EPCRA), TRI requires facilities to report their releases of certain listed chemicals to EPA, which makes the results available to the public. From 1988 to 2006 (the first and most recent reporting years), manufacturing facilities decreased their disposal or other releases by 59 percent (1.77 billion pounds) based on chemicals that have been consistently reported since 1988[16] (U.S. EPA-TRI, 2008). TRI has been so successful in the United States that other countries are developing Pollutant Release and Transfer Registers (PRTRS) including the United Kingdom, Mexico, Norway, Canada, Japan, Australia, and the Czech Republic (U.S. EPA, PRTR). Environmental Defense, an environmental advocacy organization, created "Scorecard" in 1998 to consolidate publicly available information on chemical pollutants and make it available to the public (Scorecard, 2008).

7.5   CONDUCT PERIODIC MANAGEMENT REVIEWS

Process safety performance should be reviewed by the manager who authorizes and supports the process safety and metrics systems. As mentioned above, management's commitment to act on problems identified by the metrics is as important,

> *Unless leaders respond to weaknesses identified by the metrics, collecting the data is virtually useless.*

if not more important, than the metrics themselves. Unless leaders respond to weaknesses identified by the metrics, collecting the data is virtually useless.

---

[16] The TRI program has changed over the 18-year program life with chemicals added and deleted from the list of chemicals tracked. For example, sulfuric acid was delisted in 1994 but persistent-bioaccumulative-toxic (PBT) chemicals were added in 1999. Mining and electrical utilities were included in TRI beginning in the 1998 reporting year.

Management review is the routine evaluation of whether management systems are performing as intended and producing the desired results as efficiently as possible. Such reviews should provide encouragement and praise for improvement and good performance and support and accountability for deficiencies or poor performance. It is this ongoing "due diligence" review by management that fills the gap between day-to-day work activities and periodic formal audits.

Periodic management review is a specific risk-based process safety (RBPS) element: Providing regular checkups on the health of process safety management systems—including the metrics—will identify and correct any current or incipient deficiencies before they are revealed by an audit or incident. Such management reviews also demonstrate management interest and commitment to process safety and the metrics that describe the system.

Every level of management—from the process supervisor to the facility manager to the board of directors—should conduct periodic reviews of those areas of the process safety system, including the metrics, within their spheres of responsibility. Because the objective of a management review is to spot current or incipient deficiencies in the system, such reviews should be broadly focused and ask question such as:

- What is the quality of the program?
- Are these the results we want?
- Are we working on the right things?
- What resources are needed to reach our goals by the target date?

Like in an audit, recommendations for addressing any existing or anticipated performance gaps or inefficiencies should be identified and responsibilities and schedules for addressing the recommendations assigned.

A responsibility inextricably linked to managers is the duty to provide the resources necessary to achieve objectives that are assigned to their personnel. Sometimes assigned resources are insufficient and additional resources are needed to meet expectations fully. Appropriate review and engagement by management will assure that the business case can be made and resource needs identified, or even anticipated, and provided. Promptly responding to changes, such as providing additional resources or modifying the detailed objectives, demonstrates management engagement in the systems.

Conducting management reviews demonstrates management interest in improvement and progress and reinforces a manager's stated interest in meeting objectives. Discussing results and how to make and sustain progress toward objectives reinforces commitment to achieving objectives.

7.6 CULTIVATE A POSITIVE PROCESS SAFETY CULTURE

Establishing a positive process safety culture will inspire people to support process safety and metrics efforts. Perhaps the largest enticement to implementing new or revised process safety metrics is the knowledge that the results will be taken seriously and handled appropriately, with remedial action taken when necessary. Process safety metrics provide information that can contribute to tracking the status and improvements in the organization's safety culture through elements such as resolution of findings, closed MOCs, and completion of safety training.

> *Perhaps the largest enticement to implementing new or revised process safety metrics is the knowledge that the results will be taken seriously and handled appropriately, including remedial action when necessary.*

A positive and mature safety culture promotes open dialogue on issues, discusses improvement strategies by engaging a broad array of employees, and will not tolerate deviation from established procedures. Many in an organization may not immediately internalize these seemingly "soft" concepts. It is the responsibility of management from the board of directors and senior executives downward to identify, develop, and nurture attributes of a healthy safety culture.

> *"What you do speaks so loud, I can't hear what you say."*

What management does is much more important than what it says; process safety metrics are one way of demonstrating that actions align with words.

7.6.1 Support a Continuous Improvement Culture

Visibly supporting a continuous improvement culture that benefits everyone is an effective strategy to sustain process safety performance. Improvement efforts are often cast as a program with a definite beginning and presumptive end point. Driving to zero process safety incidents will require a sustained effort with evolving focus over the life of the process operations.

As metrics identify weaknesses and deficiencies, steps must be taken to rectify them. Similarly, as metrics document performance gains, such improvements should be communicated and celebrated, but also need constant attention to maintain and/or further improve that performance. Approaching objectives with a continuous improvement mind-set allows the organization to modify objectives as changes occur and constantly raise the process safety performance bar.

7.6.2   Make Process Safety Relevant to All Personnel

A process safety program must be relevant to the company and its operations. The program must also be relevant to employees' personal safety and success. The link between safety actions and safety outcomes is most obvious in operations that directly handle highly hazardous materials. Unfortunately, that nexus may not be as obvious in the systems that support and are otherwise indirectly associated with hazardous materials. Appropriately developed, implemented, and communicated process safety metrics can help educate personnel on the importance of different aspects of the process safety system.

Case histories can bring alive not only risks but also the value of process safety as documented through metrics. Case histories must include the details of the dysfunctional behaviors and conditions in order to provide any learning benefit. As employees throughout the organization are exposed to reports of actual accident cases—whether through regular safety meetings or internal and external information sources—they will become more sensitized to the importance of the steps the organization is taking to prevent process safety incidents. Metrics reports such as near-miss reports or incident tracking help demonstrate what the organization is doing to "keep that from happening here." Most companies track, post, and publicize their occupational injury rates—"20 years without lost-time incidents," "10 consecutive years of total safety" (Air Products, 2008), and similar demonstrations of occupational safety are common in the industry. Appropriate process safety metrics will eventually allow similar acknowledgement of process safety performance.

Case histories may be drawn from inside or outside the company. Many companies internally circulate incident reports, emphasizing the reports or portions of reports that highlight particularly important or timely learnings. Such reports may be followed up with additional technical information to support corrective actions throughout the organization.

Various member and industry trade associations schedule time during meetings or otherwise provide vehicles for members to share learnings such as the API/NPRA Operating Practices Symposium and meetings sponsored by the Mary Kay O'Connor Process Safety Center (MKOCPSC, 2008) and the AIChE Global Congress on Process Safety, which includes the Center for Chemical Process Safety (CCPS). A standing item during the Contra Costa County CAER Group's quarterly Safety Summits is a sharing of lessons learned by member companies (CCC CAER, 2008).

The monthly CCPS *Process Safety Beacon* (CCPS, Beacon) is designed for manufacturing personnel, describing actual accidents, the lessons learned, and the practical means to prevent a similar accident in other plants. Similarly, the United Kingdom's Institution of Chemical Engineers (IChemE) *Loss Prevention Bulletin* (IChemE, LPB) is another major source of safety case studies for the process industries. Included with each *LPB* issue is a Powerpoint "Toolbox Talk" on one example of a type of incident or accident. IChemE has also partnered with BP to produce a series of animated process safety case studies (IChemE, ASL). Hazards Intelligence or Hint (Ility Engineering, Hint) operates an informative Web site on incidents worldwide that are collected through a network of reporters and correspondents.

The U.S. Chemical Safety and Hazard Investigation Board publishes detailed investigative reports on incidents it investigates. These reports are valuable resources for corporate and facility process safety professionals. The reports are accompanied by Investigative Digests, shortened version of the reports designed to be distributed broadly at facilities and to upper management. CSB also produces safety publications on timely process safety topics as well as videos about incidents and findings. The U.K. Health and Safety Executive (HSE) publishes Safety Alerts and other information based on its incident investigations and reports. The European Commission Joint Research Centre maintains the Major Accident Reporting System (MARS), a distributed information network consisting of 15 local databases, created to respond to the call of the Seveso II Directive for more open access to information on major accident hazards for all interested parties (MARS, 2008).

Discussing the benefits of process safety and metrics systems is an important strategy to gain employees' commitment to these systems. Reinforcing those messages by further refining the messages to relate personally to employees will help drive the commitment to process safety.

7.7   COMMUNICATE PROCESS SAFETY AND OTHER ORGANIZATIONAL SUCCESSES

Companies use various communication techniques to convey their commitment to process safety objectives as well as successes in meeting those goals (see Chapter 6). Messages should be clear and consistent, reinforcing the organization's commitment to eliminate process safety incidents. The process safety metrics become a tool to demonstrate progress and successes. Such messages can be presented in various forms and dialogues, and the common theme should always be included in communications.

Using success stories and case histories is a common technique to demonstrate the value of a process safety and metrics system. Illustrative examples are easy to understand and demonstrate the value of a reliable process safety system. Metrics provide the means to shape such stories and explain the improvement progress and benefits for all members of an organization.

Success stories demonstrate that positive change can be made within an organization. These stories provide a basis for dialogue with all personnel in an organization and facilitate education and training efforts. Metrics help document systems improvement trends and provide data to reinforce desired behaviors.

Advertising successes of the process safety improvement effort demonstrates that improvement is possible. Well-crafted stories also explain the benefits that accrue to everyone in the organization. Of particular interest are stories where a process safety weakness was observed, possibly during a process safety audit, and an improvement effort corrected the identified weakness before it could manifest into an accident. Metrics can validate such improvements. Another example is improved reliability from timely maintenance of safety devices as demonstrated by metrics that educate

> *Advertising successes of the process safety improvement effort demonstrates that improvement is possible.*

personnel not only about the hazards, but also about the importance of reliable safety systems in managing those hazards.

Success stories can show that improved process safety can contribute to increased productivity, cost efficiencies, and reliability. Using metrics to demonstrate the nexus between process safety performance and other organizational objectives only serves to reinforce individual and organizational commitments to process safety. A good process safety system, including appropriate metrics, can help prevent events such as unplanned shutdowns. Decreasing the number of unplanned shutdowns due to equipment failures or processing aberrations are examples of where process safety and other operational activities are linked. Unplanned shutdowns create extra cost and production problems as well as increase the probability of a process safety incident.

Such success stories may come from within the organization or from external sources including other companies and even other industries. Professional and trade associations, safety organizations, and some customers or suppliers as well as other business publications and sources can be sources for such information. Many of the same organizations hold meetings that discuss lessons learned (see Section 9.2.2). An organization may want to engage the corporate communications and/or public affairs functions to look for opportunities to publicize the company's successes within and outside the firm as well as look for examples outside of process safety successes.

7.8 CONCLUSION

Metrics alone will not improve process safety. Data must be collected, analyzed, communicated, understood, and *acted upon*. Management's visible responses to identified problems are as important as, if not more important than, the metrics themselves. Objectives may be defined and responsibilities assigned, yet a lack of management follow-up will undermine focus and weaken commitments. Company leadership must be committed to process safety performance. A demonstrated commitment to process safety by an organization's senior leadership is essential.

In addition, responsible parties—from the shop floor to the board of directors—should be rewarded for good process safety performance, and there should be consequences for poor performance. Linking process safety results with personnel compensation must be done carefully to avoid counterproductive incentives. The compensation systems must not discourage reporting unsafe conditions or other "gaming" of the metrics system.

Sharing performance metrics and results broadly can engage the public as a partner in holding the organization accountable for process safety performance. Making metrics and performance public can be an especially powerful way of maintaining upper management commitment since it will likely be the CEO or other senior managers who will be called to account by the public if goals are not met or performance declines. Communicating process safety successes also demonstrates to employees and the public that positive change can be, and are being, made within an organization.

Process safety performance should be reviewed by the manager who authorizes and supports the process safety and metrics systems. Unless leaders respond to weaknesses identified by the metrics, collecting the data is virtually useless.

Finally, management must cultivate a positive process safety culture that will inspire people to support process safety and metrics efforts. Perhaps the largest enticement to implementing new or revised process safety metrics is the knowledge that the results will be handled seriously and appropriately, with proper remedial action taken when necessary. A positive and mature safety culture promotes open dialogue on issues, discusses improvement strategies by engaging a broad array of employees, and will not tolerate deviation from established procedures.

REFERENCES

Air Products and Chemicals, "Environmental, Health and Safety, Our Commitment to EH&S: EH&S Awards," available at http://www.airproducts.com/Responsibility/EHS/OurCommitment/EHSAwardsSummary.htm

American Chemistry Council Responsible Care Process Safety Web site, http://www.americanchemistry.com/s_responsiblecare/sec_members.asp?CID=1318&DID=4861

BP America "Our Process Safety" Web site, a http://www.bp.com/sectiongenericarticle.do?categoryId=9023555&contentId=7040564

BP America "Texas City Investigation Reports" Web site, http://www.bp.com/genericarticle.do?categoryId=9005029&contentId=7015905 (2005)

Center for Chemical Process Safety, "Conferences and Events" Web site, http://www.aiche.org/Conferences/Specialty/GCPS.aspx

Center for Chemical Process Safety, *Process Safety Beacon* Web site, http://www.aiche.org/ccps/publications/beacon/index.aspx

Contra Costa County CAER (Community Awareness & Emergency Response) Web site, http://www.cococaer.org/prepare.html

Dow Chemical Company, "Our Commitments: Public Safety & Security: Local Protection of Human Health and the Environment," available at http://www.dow.com/commitments/goals/protect.htm

Dow Chemical Company, *Sustainability Update 2007 Year End,* May 2008

Dow Chemical Company, *2015 Sustainability Goals Update 1Q 2008,* May 2008

Dow Chemical Company, *2015 Sustainability Goals Update 2Q 2008,* August 2008

DuPont, "Safety Awards," available at http://www2.dupont.com/Consulting_ Services/en_US/news_events/leadersforum/about_awards.html

European Commission, Joint Research Centre, MARS Web site, http://mahbsrv4.jrc.it/mars/servlet/GenQuery?servletaction=ShortReports

Green Media Toolshed, "Scorecard: The Pollution Information Site," available at http://www.scorecard.org/index.tcl

Health and Safety Executive (HSE) Chemicals Manufacturer and Storage— Information Web site, http://www.hse.gov.uk/chemicals/information.htm

Hopkins, A,, "Thinking About Process Safety Indicators," *Safety Science,* Vol. 47, No. 4, 2009.

Ility Engineering, *Hazard Intelligence* Web site, *http://www.saunalahti.fi/ility/index.html*

Institute of Chemical Engineers *Animated Safety Lessons* Web site, http://cms.icheme.org/mainwebsite/general-barafc3d75d.aspx?map=5399ea 5c7cb39c90cf7337f186099afe

Institute of Chemical Engineers *Loss Prevention Bulletin* Web site, http://cms.icheme.org/mainwebsite/general-barfb7d65aa69ef5027.aspx?map= 49c04834433bef0ca60153732cf0b3be

International Council of Chemical Associations (ICCA), "ICCA Responsible Care® Status Report 2007," available at http://www.icca-chem.org/pdf/RCREPORT17-12.pdf

Mary Kay O'Connor Process Safety Center Web site, Texas A&M University, http://psc.tamu.edu

U.S. Chemical Safety & Hazard Investigation Board, Completed Investigations Web site, http://www.csb.gov/index.cfm?folder=completed_investigations&page=index

U.S. Environmental Protection Agency International Pollutant Release and Transfer Register (PRTR) Web site, *http://www.epa.gov/tri/*

U.S. Environmental Protection Agency, "Toxics Release Inventory 2006 Public Data Release," (TRI), available at http://www.epa.gov/tri/tridata/tri06/index.htm

8

IMPROVING INDUSTRY-WIDE PERFORMANCE

As individual companies commit to and improve their process safety performance, the performance of the industry as a whole will also improve. Much industry-wide improvement will come through companies' voluntary actions—benchmarking against peers, comparisons through trade associations, and learnings from industry peers. External stakeholders and regulatory bodies can also use industry performance, whether available through regulatory compliance information or other public information, to encourage continuous improvement.

An organization that measures process safety performance will likely want to know how well its performance compares to that of its industry peers. This external perspective helps the organization understand if its process safety systems are performing at generally accepted levels compared with peers within and outside the industry. Performance goals are developed internally, and an external validation of those goals is important. Organizations will find it difficult to embrace process safety goals that do not meet existing industry practices and expectations of their external stakeholders.

This chapter describes external influences upon an organization's performance objectives as well as some methods to develop and understand the external perspective. This chapter also discusses how external comparisons drive performance inside the company and across the industry.

8.1 PERFORMANCE BENCHMARKING

An outside view will provide an organization with answers to questions such as:

- How does our performance compare to our industry peers?
- Are we doing enough?
- Are we efficient?
- Are we meeting public expectations?

The answers to these questions can often be found by comparing an organization's programs and performance with those of others, or benchmarking. Benchmarking activities can range from an informal comparison of internal performance against published data to a formal, detailed performance comparison among the organization and two or more other companies or entities. Industry associations (ACC, API, NPRA), professional societies (AIChE, CCPS, IEC), academic institutions (Texas A&M), and governmental agencies (U.S. OSHA, U.S. CSB) publish a variety of statistics and other information that may be used to compare against an organization's performance. Whether through a structured exercise or a less formal process, benchmarking provides performance comparisons and the opportunity to share good practices.

Common process safety metrics, such as the number of process safety incidents, can provide a basis for such comparisons. Formal benchmarking among different companies, often facilitated by a third party, can investigate many detailed aspects of the respective process safety systems. However, comparisons will only be meaningful if the benchmarking partners use metrics with consistent definitions. The major goal of CCPS's "Process Safety Leading and Lagging Metrics" was to create such common definitions.

8.2   METRICS ALLOW PERFORMANCE COMPARISONS FOR MULTIPLE PARTIES

Common process safety metrics provide the ability to compare the performance among facilities within an organization, among different companies within an industry and even among different industries. Detailed performance metrics may vary between different companies since each company will have unique needs and improvement objectives. However, high-level process safety goals, such as the prevention of process safety incidents (as measured by the number of process safety incidents), will be consistent across the chemical and allied processing industries, and comparable progress toward such common goals can only be tracked through the adoption of common definitions and metrics.

> *Use metrics to compare process safety performance among facilities, companies, and industries.*

Even though the overarching goals may be the same, the process different organizations use to reach those goals can differ—some organizations may focus on one aspect of process safety management while another organization will focus on other aspects. While there is no single path or prescribed approach for achieving good process safety performance, there will be similar characteristics embodied in good process safety systems, and performance metrics associated with those characteristics may also be similar. If organizations use consistent metrics, benchmarking among those organizations will allow valid performance comparisons.

8.2.1   Industry Comparisons Influence Objectives

Comparing performance with industry leaders allows an organization to better understand the strengths and weaknesses in its process safety systems. An organization will often choose to emphasize improvement in areas that do not compare favorably with the performance of its industry peers. On the other hand, if the comparisons show superior performance as compared to industry peers, the organization may refocus performance improvement efforts on other, higher-priority issues.

The organization must exercise caution in changing priorities so as not to allow high-performing elements to degrade. For example, if an organization's metrics for personnel training show that training and recertification of operating staff occurs as scheduled and is effective, its practice will be considered reliable. However, as priorities shift to other process safety elements, the organization must continue its training metrics to maintain its good training performance.

Even more important, when evaluating performance against industry peers, the organization should commit to and sustain efforts toward improving or maintaining priority metrics over time. Frequently changing priorities or metric targets can confuse those trying to meet them and can potentially lead to no improvement at all. Worse yet, eventually metrics compliance could be seen as a game and not taken seriously.

Many organizations track the number of process safety incidents, and some even share that information with others. However, while a higher incident rate may indicate weakness in the overall process safety system, the incident rate alone will not identify which process safety system elements need improvement. Detailed benchmarking with industry peers can aid in developing a targeted performance improvement effort.

8.2.2   Common Definitions Lead to Successful Industry Comparisons

Comparing performance to that of industry peers will pay dividends, but proper preparation must be made before performing detailed benchmarking. A key requirement of a successful benchmarking effort is the use of common terms and definitions and similar metrics used for performance comparison. If different definitions are used to compare performance, the results will be at best confusing and at worst misleading with little likelihood of gaining valid information and insights. Adopting common metrics and definitions across a group allows data to be compared easily among the group partners. Common metrics allow not only one-on-one benchmarking, but also group benchmarking.

CCPS's "Process Safety Leading and Lagging Metrics" (CCPS, 2007b) defines process safety metrics that were developed by a broad set of chemical and allied processing industries stakeholders. The CCPS publication is a starting point to initiate or enhance, a metrics system, and it provides a foundation for consensus-based metrics. Should companies and industry associations adopt the CCPS lagging and leading metrics, more consistency will exist, providing a valid basis for performance comparison among a large number of companies. Adopting and reporting consensus metrics externally will provide transparency to outside stakeholders and build confidence in industry performance.

8.3   SHARING DATA ACROSS INDUSTRY LEADS TO IMPROVED PERFORMANCE

Readers of this book are likely seeking ideas for developing strong practices that deliver superior process safety performance in the most cost-effective manner. Sharing experiences using process safety metrics is one proven means to gain the desired knowledge. As metrics are accumulated and data is shared among multiple stakeholders, a dialogue that identifies strong practices ensues. As strong practices are identified and good performance is validated, these practices often become the norm. Using uniformly strong practices across an industry group will improve the overall group process safety performance. As industry associations recognize the success of strong practices, they will encourage the industry association members to adopt those practices.

8.3.1   Regulatory Mandated Data Can Facilitate Comparisons

Reporting mandated by regulation can lead to improved performance if the organization uses regulatory reported information as metrics to recognize strengths and weaknesses. Countries around the world have defined criteria for reporting various process safety events, such as unintended material releases that exceed specific reporting thresholds, workers' injuries, and process safety incidents. Obviously, a very high number of reported incidents will catch the attention of regulatory agencies, leading to increased pressure to improve performance.

8.3.2   Incident Investigations Reported by Others Can Provide Valuable Insights

The experience of those that suffer process safety incidents is invaluable information, and learning from the unfortunate experiences of others can help an organization select appropriate metrics. As companies learn from their incidents, they often share this knowledge with other companies and industry peers. Companies should look for opportunities to tap into this external information source.

8.3.2.1 Industry Associations, Professional Societies, and Government Agencies Provide Valuable Information

Industry associations, professional societies, academic institutions, and government agencies provide information on process safety incidents. Several groups sponsor conferences (both domestic and international) that include major sessions on process safety incidents. Documented proceedings from these conferences provide valuable information for selecting metrics based upon historical incidents. The U.S. Chemical Safety and Hazards Investigation Board (CSB) conducts incident investigations and hazards studies, and documents causal factors and other causal information on incidents. The U.K. HSE publishes guidance on a variety of subjects (Chapter 4 discusses HSG254, an HSE publication on metrics).

8.3.2.2 Incident Investigations by Others Identify Weakness in Defined Metrics

After an incident occurs, questions are often asked: "Why didn't we see this coming?" or "What are the factors that resulted in our not finding the causal factors before the event?" Incident investigation results not only provide insights as to how the process safety system may have failed, but also provide information on how well the existing metrics may have foretold weaknesses leading up to the incident. The experience of other organizations can help determine the effectiveness of certain metrics as companies ask themselves, "Could this event or a similar one occur in one of our operations?" Externally reported incident investigation learnings can be used to determine if modifying metrics may result in more reliable prediction of systematic failures.

> *Industrial and professional associations provide forums to discuss process safety incidents learnings. Examples include:*
>
> - *AIChE – Ethylene Producers Conference*
> - *AIChE – Global Congress on Process Safety*
> - *Ammonia Institute Symposium*
> - *EFCE Symposia on Loss Prevention and Safety Promotion in the Process Industries*
> - *IChemE Loss Prevention Conference*
> - *Mary Kay O'Connor Process Safety Center Conference*
> - *NPRA National Safety Conference*
> - *NPRA/API Operating Practices Symposium*
> - *US CSB*

8.3.3   Shared Performance Metrics Can Spark Improvement

When organizations track their process safety performance, they usually develop a desire to improve performance if there is available data that indicates improvement is possible. Recognizing that industry peers have demonstrated better performance will provide motivation to improve performance. It is commonly held that avoiding process safety incidents will benefit any organization, and catastrophic incidents are clearly unacceptable. Since a higher frequency of process safety incidents is usually associated with weaknesses in the process safety systems, a high incident frequency reported to an industrial association should command attention from upper management. Metrics provide the means to evaluate just how well an organization is doing in managing process safety. Discussion of process safety performance with peers and others in the processing industries is strong reinforcement for advocating reliable process safety performance. A company's comparison to industry peers' performance allows the organization to understand if its performance is meeting expectations.

8.3.4   Industry Associations Track Members' Performance

Some industry associations periodically collect from their members data on the number of process safety incidents. This data is often a high-level metrics, such as the number of process safety incidents, that is useful in describing the overall process safety performance for the association membership. The metrics are normally expected to illustrate an improvement trend and support their overarching goals, such as eliminating process safety incidents. Year-to-year comparisons can be made to demonstrate if the trend is improving (decreasing number of incidents). This data can be used to compare performance among members; however, it is not particularly useful in determining improvement opportunities for specific process safety elements. The number of process safety incidents is tracked, but depending

> *Industry associations will often collect process safety performance data, including:*
>
> * *The American Chemistry Council*
>
> * *The American Petroleum Institute*
>
> * *The Chlorine Institute*
>
> * *The National Petrochemical Refiners Association*
>
> * *The Texas Chemical Council*

upon the severity threshold for reporting such incidents (few high-consequence incidents occur in a given year), such data may not be specific enough for detailed systems evaluation.

8.3.4.1 Compare Process Safety Performance With Others

An industry association that collects process safety metrics allows its members to compare their performance to that of other members. In some cases the reported data does not allow for identification of individual members' performance, so the comparisons can only be made to the association's reported norms. In other cases, more detailed information may be reported. In either case, this information can be valuable for validating a company's objectives and improvement efforts to ensure the company does not lag behind industry performance.

Members of an association or organization may also use such data to benchmark against competitors. Similarly, subsets within an organization may use the data to compare performance within the subset as well as against the broader industry performance. For example, pharmaceutical companies or refinery members of CCPS may compare performance against others in their subsector as well as broader process industry performance.

8.3.4.2 Management Uses Industry Association Performance Data

Upper management in many companies uses industry association metrics as a means to quickly compare their organization's performance to others in their peer group. This information is used to decide if their performance is sufficient given the performance of others in the same or similar industry.

8.3.5 Managers' Self-Esteem Often Drives Performance

Most company executives do not want to be seen as managing their business in a subpar fashion, and process safety is one business risk that upper management cannot afford to manage poorly. Discussions among industry peers help form a self-assessment for process safety performance based upon norms established by industry members. Self-esteem will often energize company management to drive process safety performance up to, or beyond, the industry norm. Management in some companies will purposely set challenging goals in order to be recognized as performance leaders within their industry.

Management in some companies will purposely set goals so as to be recognized performance leaders within their industry. Such goals are often linked to a strong desire to make the public aware of the company's commitment to high performance in all aspects of its operations. Often these companies allow the public to track and hold the company accountable for meeting performance goals. Defining a goal to be a performance leader in process safety is similar to being a performance leader in environmental and health performance, and this is often consistent with a company's documented values.

8.3.6 External Stakeholder Use Metrics for Performance Comparisons

External stakeholders for an organization, such as stockholders, advocacy organizations, and the public, expect indications of how a company is performing in all aspects of its operations. Comparing metrics across a processing industry segment allows external stakeholders to understand if performance is meeting expectations. If expectations are not met, pressure from external stakeholders will be applied for improvement. Indicators of good performance provide a positive discussion point for the company and external stakeholders. Fatalities per year across all industry segments have been tracked by regulators and other interested parties for years. These metrics are publicly available, and external stakeholders in many cases have successfully pressured poor performers to improve.

8.3.7 Regulators May Use Industry Performance Data

Regulators will have access to data published by industrial associations. This information allows regulators to supplement their data and to make decisions on how to focus their efforts to improve process safety in the chemical and allied processing industries. Access to such information will allow regulators to develop a better understanding of issues facing the processing industries. External stakeholders and regulators may use an organization's metrics to unfairly persecute the organization. This cannot be prevented, but should not be allowed to adversely influence decisions on metrics.

8.4 CONCLUSION

An organization that measures process safety performance will want to know how well its performance compares to that of industry peers. As individual companies improve their process safety performance, the performance of the industry as a whole will also improve. One way of improving individual performance is to compare it with similar performances of others.

Common process safety metrics, such as the number of process safety incidents, can provide a basis for such comparisons among facilities within an organization, among different companies within an industry, and even among different industries. The major goal of CCPS's "Process Safety Leading and Lagging Metrics" is to create such common definitions.

Much industry-wide improvement will come through actions such as benchmarking against peers, comparisons through trade associations, and learnings from industry peers. Benchmarking, within an organization or against industry peers, provides performance comparisons and the opportunity to share good practices, whether conducted through a structured exercise or a less-formal process.

Sharing process safety performance data and experience through an industry trade or professional organization can be a valuable and cost-effective source of valuable information on a company's performance measured against industry peers. Whether through the sharing of actual data or an ongoing dialogue among peers, strong practices will be identified and good performance is validated and often become the industry norms.

Comparison with industry peers can be a strong motivator for improved performance, especially when industry peers have demonstrated better performance is possible. This is true from the shop floor to the executive offices. Company executives do not want to be seen as managing their business in a subpar fashion, and process safety is one business risk that upper management cannot afford to manage poorly.

External stakeholders such as stockholders, advocacy organizations, and the community can and do use regulatory compliance and industry-provided performance data to evaluate the performance of individual facilities, companies, and whole industries. Stakeholders hold the industry accountable for process safety performance and, when they perceive performance to be lacking, will use the information to advocate improvement. Regulators may also use such information to make decisions on how to focus their enforcement and regulatory efforts.

REFERENCES

Center for Chemical Process Safety, "Process Safety Leading and Lagging Metrics," American Institute of Chemical Engineers, New York, 2007

9

FUTURE TRENDS IN THE DEVELOPMENT AND USE OF PROCESS SAFETY METRICS

Metrics will become increasingly important as the discipline of process safety continues to evolve and improve. They hold promise to contribute to risk-based process safety, inherent safety, safety culture, and compliance with regulation as well as broader societal interests surrounding sustainability, globalization, and societal risk aversion.

9.1 IMPROVING PROCESS SAFETY

9.1.1 Consensus Metrics

As discussed earlier in the book, common metrics are necessary for companies to compare their performance with other companies and overall industry performance. Consensus process safety metrics are only beginning to be adopted within the processing industries. Consensus process safety metrics are being developed, such as those recommended in the CCPS publication "Process Safety Leading and Lagging Metrics." Pressure from within the industries and from outside stakeholders will encourage the broader acceptance of current consensus metrics as well as the development of more such metrics. Such metrics will be used not only by companies and industries, but especially by outside parties to evaluate industry-wide performance or the performance of individual companies against the industry as a whole.

9.1.2 Continuous Improvement

Establishing process safety metrics is not a one-time exercise. They are not cast in stone, never to be revisited. Instead, metrics should be reviewed and revised or even retired as part of the company's ongoing commitment to continuous improvement. One way this can be done is to revisit metrics on a regular schedule when a particular section of the facility or company process safety procedures are reviewed or when a particular metric is consistently met. A revision point allows a consideration of whether it may be possible to "deepen" the metric, to look for modifications that would take performance to the next level.

For example, a metric on the total number of safety incidents may lose its effectiveness in driving improvements as an organization approaches the target of zero incidents. In fact, it may drive the wrong behavior, e.g., moving away from collaborating on solutions to simply blaming the unit where the incident occurred. This signals the need for new metrics. Metrics on the total number of incidents may be replaced by process metrics reflecting the causes behind the incidents, and thus provide more in-depth analysis. They serve as the new set of metrics on which the organization focuses (GEMI, 2007). Another example might be to evolve from tracking PHAs completed on schedule to evaluating the quality of the PHA or the appropriateness of the technique(s) used to identify hazards and even tracking how the results of PHAs are reflected in operating procedures, training, mechanical integrity programs, or other areas that might benefit from reflecting PHA learnings.

9.1.3   Risk-Based Process Safety

In 2007, the Center for Chemical Process Safety (CCPS) published *Guidelines for Risk Based Process Safety* (CCPS, 2007a) to help organizations design and implement more effective process safety management systems, thereby halting a trend toward stagnant process safety management and possible serious decline in process safety performance. The lack of widely recognized consistent measurement systems available for process safety was identified as one of the possible causes of process safety management performance stagnation.

A risk-based process safety (RBPS) management system relies on a combination of leading and lagging indicators to provide a complete picture of process safety effectiveness. One of the major challenges in moving toward RBPS is managing the initial difficulty in selecting the appropriate performance metrics and acquiring the discipline and measurement systems required to maintain these performance metrics. Yet understanding hazards and risk, monitoring an appropriate suite of leading and lagging indicators, and periodically conducting management reviews are critical to highlighting strengths, identifying weaknesses, and taking corrective action in a timely manner. Appendix I summarizes the types of leading and lagging metrics an organization may want to consider in implementing risk-based process safety.

9.1.4   Inherent Safety

Inherent safety (IS) is defined by the Center for Chemical Process Safety as "a concept, an approach to safety that focuses on eliminating or reducing the hazards associated with a set of conditions" (CCPS, 2008). Many process safety professionals have voluntarily embraced inherent safety concepts and methods for addressing process risks. Ideally, all companies will apply inherent safety principles throughout the design and operating life cycle of projects (CCPS, 2008). Regulators and legislators at the national, state, and even local levels have likewise recognized the risk reduction potential in IS and have debated whether they could improve overall safety or security results by encouraging or mandating IS approaches through regulation and, if so, the best way to construct such regulations.

IS opportunities may exist throughout the life cycle of a process, but specific IS design and implementation is a function of site and process conditions. Such site/process specific options may be less conducive to regulation or objective comparison. Both industry entities and regulators lack tools and measures to compare the inherent safety of multiple options or to determine what is "feasible." As organizations develop programs to implement IS concepts through the life cycle of processes, the process safety metrics program can be used to track implementation of the IS program as well as program benefits.

9.1.5 Quality Assurance/Quality Control During Detailed Design, Construction, and Maintenance

Poorly detailed design or poor construction of a good design can create additional hazards in a system, such as incorrectly sized relief systems or incorrect construction materials. A mature process safety metrics program can be used to track the quality of process design, construction, and maintenance. A disciplined quality assurance/quality control (QA/QC) process for design and construction can help prevent such hazards. Similarly, hazards may be introduced if a QA/QC process is not part of the maintenance operations. Detailed design, construction, and maintenance metrics have traditionally focused on cost and schedule. A process safety metrics program could be expanded to include design, construction, and maintenance QA/QC elements to reduce the likelihood that such hazards are created or introduced. Alternatively, such QA/QC elements could be integrated into an inherent safety program that itself is monitored as part of a mature process safety metrics program.

9.1.6 Reliability

CCPS defines reliability as the probability that an item is able to perform a required function under stated conditions for a stated period of time or for a stated demand. In addition to relying on QA/QC, companies use dependable data to conduct reliability analyses. Though related, reliability is different from quality. While quality control is concerned with the performance of a product or process at one time, reliability is concerned with the performance of a product over its entire lifetime. Reliability engineering addresses all aspects of a product's life, from its conception, subsequent design, and production processes, through its practical use lifetime, with maintenance support and availability, and covers reliability, maintainability, and availability. Process safety metrics data can provide valuable input to the life data analyses used to estimate the probability and capability of parts, components, and systems to perform their required functions for desired periods without failure, in specified environments.

9.1.7 Process Safety Culture

An emphasis on process safety culture will continue to be a key element in maintaining and improving process safety performance. This is especially true in an increasingly competitive global business environment where maintaining an organization's fundamental safety commitment at all its locations becomes increasingly difficult because of challenges such as cost pressures, geographical separation, downsizing, and reorganizations. The strength of the organization's process safety culture—the shared values, beliefs, and perceptions—will help determine how the organization responds to such challenges. The values that underlie the organization's process safety culture will help individuals to understand, accept, and do what is right even when no written rules or procedures are in place to address a particular situation.

While the behaviors and attitudes associated with a sound culture, once established, should become the norm for an organization, staff turnover will continually require the instilling of group values and attitudes in new members of the organization. External factors, such as economic pressures, may have a potentially erosive effect on the culture. Thus, while the intensity of effort required for sustaining a culture may not be as great as that required to effect a step change, a continuing effort will nevertheless be required. Developing and maintaining a vibrant process safety culture is a long-term effort that will require dedicated resources for as long as the organization exists.

CCPS has identified 11 key attributes of a sound safety culture[17] (CCPS, 2005) and process safety culture is a key element in risk-based process safety. Increasingly, organizations will use culture surveys or surveys in conjunction with other outreach techniques to monitor process safety culture. Surveys should be anonymous and survey participants should be encouraged to be as open and honest as possible. Survey data should then be analyzed and the results of the analysis given to the workforce. Key areas for improvement should be identified and workforce engagement should be sought around their ideas of how to improve. Workforce engagement is critical in moving towards any improvement plan. It is expected that companies will increasingly develop and implement programs to enhance safety/operating culture and this type of methodology will be used to track the effectiveness of such programs

---

[17] The list of safety culture attributes is based upon the content of an ABS Consulting continuing education course, *Creating and Sustaining a Sound Safety Culture*, and is used with the permission of ABS Consulting: (1) Espouse safety as a core value; (2) Provide strong leadership; (3) Establish and enforce high standards of performance; (4) Maintain a sense of vulnerability; (5) Empower individuals to successfully fulfill their safety responsibilities; (6) Provide deference to expertise; (7) Ensure open and effective communications; (8) Establish a questioning/learning environment; (9) Foster mutual trust; (10) Provide timely response to safety issues and concerns; and (11) Provide continuous monitoring of performance.

9.1.8 Enhanced Databases and Dashboards

As discussed in Chapter 6, dashboards will increasingly be used to communicate process safety metrics and performance to local line management. Such dashboards may reflect current, summary, or cumulative data (such as the number of days without a leak), current conditions (such as the fill level of a tank), or the results of the most current inspection or audit. An organization's process safety metrics can evolve in a hierarchical way so that individual indicators can be rolled up from the process level to the site level and eventually even the organizational level. Dashboards can be expected to evolve similarly to report such "rolled up" information to upper management.

Companies can be expected to rely increasingly on online databases, populated by the operating plants that can be integrated to generate aggregated process safety performance reports. Certain standard reports can be generated automatically based on the needs of the specific audiences. Such reports may also be accessible from a dashboard. As organizations become increasingly reliant on dashboards to access process safety information, the organization may even develop measures to reflect the quality of process safety elements, such as a scale that ranges from minimum, basic performance elements to proactive, deeply embedded, mature, and high-functioning performance. Such a representation could be particularly useful to facilities or organizations that have embarked on a significant upgrade to their process safety management system, such as implementation of risk-based process safety. Since organizations will likely choose to implement different RBPS elements over time, and a dashboard could reflect the implementation stage of the different elements at different facilities within the organization or different processes within a facility.

An example is the public Web site maintained by the U.S. Nuclear Regulatory Commission (NRC) as part of its reactor oversight program (ROP). The site provides both summary and individual plant performance information for 16 performance indicators in the areas of reactor safety, radiation safety, and safeguards (NRC). The current quarterly performance indicators and assessment of inspection findings are provided using a color notation of significance—green, white, yellow, or red—and a link to the statistics and inspection findings underlying the color. The inspection and performance indicator information is also presented in summary form that allows comparison across all the nuclear plants.

While the use of centralized data management systems and dashboard displays are generally for internal purposes, companies and organization may one day develop public dashboards that would allow external stakeholders to track performance.

9.1.9 Regulatory Prioritization

Regulatory agencies—whether national or more local— do not have the resources to inspect every facility every year. As a result, regulatory agencies may schedule inspections based on available information and criteria. Agencies and the regulated industries have seen merit in using statistics on occupational injuries to monitor improvement and to target intervention. However, as discussed earlier, these metrics are insufficient and inappropriate for predicting a facility's risk for a catastrophic process safety event. Even reliance on historic accident rates by industry categories or group is only a very rough targeting tool. If made available to regulators, improved process safety metrics could assist agencies responsible for reducing the likelihood of low-frequency/high-impact events to give credit for companies and facilities with good performance and better target inspections and enforcement to those companies and facilities that may benefit from closer attention. Facilities might elect to voluntarily share such information with regulators to justify differential regulatory treatment, similar to the current OSHA Voluntary Protection Programs (VPPs) which provide relief to facilities that have implemented comprehensive safety and health management systems.

Currently regulatory agencies and many companies rely on factors such as determining if the elements of a process safety management program are in place; evaluating the programs for compliance with the requirements of the standard; and verifying compliance with the standard through interviews, data sampling, and field observations to evaluate process safety systems. These factors can confirm the existence of the elements of the process safety management system or program but lack the ability to estimate how well the program or system is actually working. As with regulatory targeting, improved metrics—especially leading metrics—may also improve the quality and value of both company and regulatory audits and inspections by providing better criteria against which to judge the actual safety of a facility.

9.2   SOCIETAL INTERESTS

9.2.1   Sustainability

Sustainability means "meeting the needs of the present without compromising the ability of future generations to meet their own needs."[18] Chemical and allied processing industries are being challenged to report progress toward achieving greater sustainability in the areas of societal and environmental as well as economic performance, the "triple bottom line."[19] Companies are being asked not just to provide goods and services, but assess and report on *what* they produce and *how* they produce it. Businesses' sustainability case studies have tended to

---

[18] United Nations, "Report of the World Commission on Environment and Development," General Assembly Resolution 42/187, December 11, 1987.

[19] Triple bottom line (TBL) refers to expanding the traditional business-reporting framework to take into account environmental and social performance in addition to financial performance. The TBL concept postulates that a company's responsibility is to "stakeholders" rather than shareholders. In this case, "stakeholders" refers to anyone who is influenced, either directly or indirectly, by the actions of the firm.

emphasize progress in environmental (emissions reduction, "greener" products and production) and occupational safety (reduced OII reports) rather than improvements in process safety. As organizations incorporate more holistic approaches to managing all risks, they can be expected to evolve from meeting minimum standards to moving beyond compliance through developing internal policies and/or adhering to voluntary standards to eventually becoming an industry leader (GEMI, 2007).

The Ethical Investment Research Services (EIRIS) and Sustainable Asset Management (SAM) have developed and applied composite indices to assess corporate sustainability practices based on company questionnaires and analysis of public information. The Investor Responsibility Research Center (IRRC) pointed out the value-chain perspective that analysts increasingly apply, considering risks beyond the company's immediate sphere of control.

Process safety—and the metrics to identify and track it—can be incorporated into a company's overall sustainability program. Adoption of improved process safety leading and lagging indicators that document improving process safety performance will be important in maintaining the organization's "license to operate" broadly as well as in a specific location.

9.2.2   Public Transparency

Corporate transparency—the openness of an organization with regard to sharing information about how it operates—has emerged as a focal point of societal expectations. Increasingly, corporations are experiencing pressures from stakeholders to be more transparent about their values, commitments, and performance. In this "show me" world, stakeholders want to know who the company is, what it stands for, where it is going, and whether it is living up to its commitments to society. Companies have learned—at times the hard way—that their license to operate, more and more, depends on having the public's trust (GEMI, 2004). Companies can expect to make more process safety information available to the public, whether as part of a voluntary corporate performance communications program, new government requirements, or due to increased public demand.

Companies, both individually and through industry organizations and trade associations, will publicly report performance against requirements. For example, the American Chemistry Council maintains a public Web site that tracks Responsible Care® performance by individual member company and membership aggregate. The Business Roundtable annually reports on its members programs and progress in a variety of sustainability projects through its S.E.E. Change initiative. Increasingly companies are establishing and maintaining a two-way dialogue between the organization and stakeholders. Many companies are already supporting initiatives that not only make more information on performance and societal initiatives available, but also solicit questions and feedback from the public.

Government regulators also will increase the amount of information available on the regulated community's performance. For example, the U. S. Environmental Protection Agency's Envirofacts Data Warehouse contains information on facilities subject to environmental regulation, including facility response plans for handling an oil spill. Regulators will continue to make compliance and enforcement information more easily available to and usable by the public.

9.2.3   Financial Analysis and Evaluation

There is growing interest by investors to invest in firms and funds that reflect the investors' own social values. These funds are tending to focus on the more immediate sustainability factors such as carbon footprint, energy efficiency, and environmental performance. As analysts become more familiar with process safety metrics—and those metrics are demonstrated to be useful predictors of a company's future safety performance—it is reasonable that a similar value could be placed on firms that are demonstrating their commitment to process safety as a separate factor or process safety as an element of other social values. Such demonstrated commitment to safety performance should also provide additional information, above and beyond financial information, on which the analyst can assess the overall risk of a firm or facility.

9.2.4   Financing/Insurance

One argument for quality management and management systems has been the belief that financial institutions and insurers would reflect outstanding performance in the area of safety, health, and environment with financial benefits such as lower interest rates or lower insurance premiums. While some companies have been able to negotiate such financial benefits, for the most part such hopes have gone unrealized. However, if process safety metrics become accepted as good predictors of a company's expected safety performance—particularly as predictors of low-frequency/high-impact (including high-cost) events—the financial and insurance industries may start providing financial incentives to companies with exemplary process safety performance and risk.

The reverse would be that insurance and other financial institutions would require demonstration of higher safety performance as a prerequisite for financing or underwriting risks. For example, in order to operate, nuclear power plants in the United States must demonstrate sufficient financial assurance to satisfy liability claims of members of the public for personal injury and property damage in the event of a catastrophic nuclear accident. Such insurance is available through the American Nuclear Insurers, which evaluates plant performance against objectives, criteria, and guidelines developed in conjunction with the U.S. NRC and the Institute of Nuclear Power Operations. A plant must receive a favorable report from this review before it can be insured.

9.2.5   Globalization

Increasing globalization in the processing industries creates a need for consistent metrics within global companies that will be acceptable worldwide. Management systems applicable to global operations rather than different systems in different locations could not only save resources for a global company, but allow much better understanding and comparisons of the business. Such global companies will benefit as new process safety metrics evolve and are accepted worldwide.

Increased global interaction and communications among governments and pro-active citizens around the world also creates incentive for the development of process safety metrics that will be acceptable and understood globally. Activists will continue to pressure governments and the chemical and allied processing industry to demonstrate a high commitment to safety—and provide information to back up that commitment—wherever they operate.

The Strategic Approach to International Chemicals Management (SAICM)[20] establishes a policy framework for international action on chemical hazards, especially in developing countries and countries with economies in transition. The SAICM includes 273 identified activities, including the development of "integrated national and international systems to prevent major industrial accidents and for emergency preparedness and response to all accidents and natural disasters involving chemicals" to be implemented through, among other vehicles, the International Labour Organization (ILO) Conventions No. 174 on the Prevention of Major Industrial Accidents and the OECD Project on Safety Performance Indicators. Progress in the activity will be measured based on the establishment and implementation of "integrated systems and centres to prevent major industrial accidents" (UNEP).

9.2.6   Risk Aversion

Worldwide, society is increasingly risk-averse and demanding more information and improved performance from economic entities perceived as high risk. Governments and citizens around the world will expect and require an ever-higher level of process safety performance and a demonstrated commitment to continuous performance improvement. A commitment to process safety borne out by relevant and effective metrics will increasingly become part of the license to operate everywhere.

---

[20] The SAICM operates under the auspices of the United Nations Environment Programme (UNEP) and promotes international action on chemical hazards. The SAICM was adopted by the International Conference on Chemicals Management (ICCM) in 2006 to further the goal that chemicals are produced and used in ways that minimize significant adverse impacts on the environment and human health by 2020.

REFERENCES

American Chemistry Council Web site, http://www.americanchemistry.com/ s_responsiblecare/doc.asp?CID=1298&DID=5084

American Nuclear Insurers Web site, http://www.nuclearinsurance.com/ Insurance.html

Business Roundtable, "'SEE'ing Change: 2008 Progress Report," Washington, DC, 2008, available at http://seechange.businessroundtable.org/

Center for Chemical Process Safety, *Guidelines for Risk Based Process Safety*, American Institute of Chemical Engineers, New York, 2007

Center for Chemical Process Safety, *Inherently Safer Chemical Processes: A Life Cycle Approach, 2nd Edition*, American Institute of Chemical Engineers, New York, 2008

Center for Chemical Process Safety, "Process Safety Leading and Lagging Metrics," American Institute of Chemical Engineers, New York, 2007

Elkington, John, *Cannibals with Forks: The Triple Bottom Line of 21st Century Business*, New Society Publishers, Gabriola Island BC, Canada, 1998

Global Environmental Management Initiative (GEMI), "The Metrics Navigator," Washington, DC, 2007, available at www.gemi.org

Global Environmental Management Initiative (GEMI), "Transparency: A Path to Public Trust," Washington, DC, 2004, available at www.gemi.org

Institute of Nuclear Power Operations Web site, http://www.inpo.info/

Interview, Isadore "Irv" Rosenthal, former board member of the U.S. Chemical Safety and Hazard Investigation Board, member of the BP U.S. Refineries Independent Safety Review Panel (the Baker Panel), and Senior Research Fellow at the Wharton Risk Management and Decision Processes Center, 2008

United Nations, "Report of the World Commission on Environment and Development," General Assembly Resolution 42/187, 11 December 1987

United Nations Environment Programme (UNEP), "Strategic Approach to International Chemicals Management: Comprising the Dubai Declaration on International Chemicals Management, the Overarching Policy Strategy and the Global Plan of Action," June 6, 2006

U.S. Environmental Protection Agency Envirofacts Data Warehouse Web site, http://www.epa.gov/enviro/

U.S. Nuclear Regulatory Commission, "Fact Sheet: Nuclear Insurance and Disaster Relief Funds," February 2008, available at *http://www.nrc.gov/ reading-rm/doc-collections/fact-sheets/funds-fs.pdf*

U.S. Nuclear Regulatory Commission Reactor Oversight Process Web site, http://www.nrc.gov/NRR/OVERSIGHT/ASSESS/index.html

U.S. Occupational Safety and Health Administration Voluntary Protection Programs (VPP) Web site, http://www.osha.gov/dcsp/vpp/

APPENDIX I: LISTING OF POTENTIAL PROCESS SAFETY METRICS TO CONSIDER (BASED ON THE RISK BASED PROCESS SAFETY ELEMENTS)

CCPS' *Guidelines for Risk Based Process Safety* presents a range of metrics for each of the RBPS elements. Each element includes examples for "Maintaining a Dependable Practice" as well as element-specific suggestions. These metrics are presented as examples, not recommendations or expectations. Practitioners may select from this list or use the list as a point of departure for developing metrics more applicable to their situation. This is a summary listing only; for the complete list, see the companion CD.

Practitioners are encouraged to look for opportunities to implement RBPS elements as opportunities and needs arise. Many of these metrics are shown as absolute numbers, while others are shown as percentages. In most cases, the metric can be formatted either way and converted from one to another.

PROCESS SAFETY CULTURE	Develop & Implement a Sound Culture	Monitor & Guide the Culture	Maintain a Dependable Practice
Results of periodic employee attitude or perception surveys			
Number of open recommendations (from risk analyses, incident investigations, audits, safety suggestions)	X		
Frequency with which upper managers visit the worksite, or percentage of the scheduled visits that actually take place			X
Number of process safety metrics frequency with which they are communicated with the executive leadership team (or board of directors)		X	
Percentage of near misses and incidents identified as being caused by unsafe acts or shortcuts	X		
Percentage of the required attendance achieved for meetings addressing process safety	X		
Average response time to the resolution of a process safety suggestion	X		
Number of meetings addressing process safety that are conducted per year	X		
Frequency with which relevant process safety statistics are shared with the organization	X		
Number of process safety suggestions reported each month	X		
Percentage of managers and supervisors trained on the importance of, and approaches to create and reinforce, a sound process safety culture			X
Relative frequency and emphasis of process safety–related topics in other topics such as cost, quality, and production in management communications			X
Number of communications from senior management to the general workforce promoting/discussing process safety			X
Number of times per month managers (area/unit) visit worksites and discuss process safety			X
Manager attendance at management review meetings	X		

COMPLIANCE WITH STANDARDS	Conduct Compliance Work Activities	Follow Through on Decisions, Actions, and Use of Compliance Results	Maintain a Dependable Practice
Number of incidents/near misses related to standard violations			X
Number of compliance violations per year		X	
Number of nonconformances to nonregulatory standards per audit		X	
Percentage of existing standards reviewed to ensure evergreen status (per a pre-defined schedule each year)	X		
Average amount of calendar time between standards system review completion & closeout of all action items		X	
Number of people trained on standards activities			X
Percentage of training on standards completed per schedule			X
Number of existing standards revised per year	X		
Average amount of calendar time taken for standards reviews			X
Number of standards organizations meetings attended per year	X		
Number of new sources of standards identified and adopted during the past year			X
Number of audits in which standards element personnel participated	X		
Number of identified standards applicability changes	X		

WORKFORCE INVOLVEMENT	Conduct Work Activities	Monitor the System for Effectiveness	Actively Promote the Workforce Involvement Program	Maintain a Dependable Practice
Percentage of workers who have participated in key defined workforce involvement activities, such as submitting a suggestion, serving on a risk analysis team, or participating in an investigation over the last 12 months	X			
Number of accepted suggestions that have not been implemented and average/maximum delinquency		X		
Results of worker attitude surveys with respect to acceptance of process safety responsibilities			X	
Number of suggestions that have not been evaluated (no decision made to accept or reject) and average/max delinquency		X		
Rate of submittal of worker suggestions, and changes in rate with time	X			
Percentage of suggestions accepted		X		
Percentage of personnel having the process safety aspects of their job defined in writing				X

PROCESS SAFETY COMPETENCY	Execute Activities that Help Maintain and Enhance Process Safety	Evaluate and Share Results	Adjust Plans	Maintain a Dependable Practice
Number of incidents with root cause of "insufficient process training/knowledge"	X			
Percentage of process safety training completed vs. plan		X		
Frequency with which incident investigation teams determine that the basic physical or chemical phenomenon that caused an incident was not known within the organization	X			
Presence of objectives related to enhancing process safety competence in each manager's, supervisor's, and technical staff member's personal performance plans, and in the trend over time				X
Cumulative number of days (or weeks) that key process safety positions have gone unstaffed, per unit of time (e.g., calculate as unstaffed weeks per month or per quarter)				X
Ratio of process changes to updates to the technology manual	X			
Opinion surveys on how technical information is stored and the efficacy of searches	X			
Frequency with which incidents recur because the organization has allowed safeguards that were implemented as a result	X			
Number (or percentage) of technology steward positions that are currently staffed	X			
Use of the technology manual ("hits" if it is implemented in a web-based fashion)	X			
Average response time for questions posed to the technology steward or center of excellence	X			

GUIDELINES FOR PROCESS SAFETY METRICS

PROCESS SAFETY COMPETENCY	Execute Activities that Help Maintain and Enhance Process Safety	Evaluate and Share Results	Adjust Plans	Maintain a Dependable Practice
Staff-hours (per year) devoted by the technology steward to supporting troubleshooting efforts for/with each operating unit	X			
Number of questions posed to the center of excellence per unit time	X			
Opinion surveys about the organization's competence, including opinions on the trend over time		X		
Comparison of actual to budgeted spending for activities associated with execution of the learning plan				X
Related to section 1, #2: percentage of positions key to process safety that are currently staffed				X
Related to section 1, #2: percentage of positions key to process safety that are staffed by personnel meeting experience requirements				X
Related to section 1, #2: percentage of positions key to process safety with trained/qualified backfills in place				X
Number of training days for personnel transitioning into key process safety–related positions				X
Staff-hours (per year) devoted by the technology steward to face-to-face contact with operating units	X			
Number of facilities or business units within the company that maintain up-to-date succession plans	X			
Pareto analysis of the topics discussed with the technology steward or with centers of excellence, or researched using the technology manual (assuming it is web-based)	X			
Opinion surveys on the usefulness of each center of excellence		X		

PROCESS SAFETY COMPETENCY	Execute Activities that Help Maintain and Enhance Process Safety	Evaluate and Share Results	Adjust Plans	Maintain a Dependable Practice
Surveys on the usefulness of attending technical meetings			X	
Related to section 1, #2: percentage of positions key to process safety with incoming personnel trained upon assuming position				X
Opinion surveys regarding the effectiveness of programs to promote learning				X

STAKEHOLDER OUTREACH	Identify Communication and Outreach Needs	Conduct Communication/Outreach Activities	Follow Through on Commitments and Actions	Maintain a Dependable Practice
Number of complaints received by the facility		X		X
Number of community advisory panel (CAP) meetings and attendance rate		X		
Community attitude survey results		X		X
Number of community members attending planned outreach functions such as plant tours and open houses		X		
Length of time to respond to community inquiries			X	
Number of CAP members that choose to stay involved	X			X
Number of information requests from the community		X		
Number of outreach activities per month/year		X		
Annual number of inquiries from regulators	X			
Number of industry group meetings at which company presenters shared significant lessons learned		X		
Cost incurred for attendance at industry group meetings		X		
Number of management review meetings held to discuss outreach issues			X	
Requests granted by regulators				X
Cost associated with regulatory citations				X
Number of community members participating in an annual "open house"		X		
Frequency of review or update public/community communications (e.g., pamphlet, video)			X	
Number of new stakeholders identified	X			

STAKEHOLDER OUTREACH	Identify Communication and Outreach Needs	Conduct Communication/Outreach Activities	Follow Through on Commitments and Actions	Maintain a Dependable Practice
Number of new/revised communications plans	X			
Number of key personnel who have received initial or refresher communications, outreach, or crisis management training		X		
Number of press briefings on the company		X		
Number of outreach meetings held		X		
Amount of time spent preparing for and conducting CAP meetings		X		
Number of commitments made to the community versus the number of commitments completed			X	
Cost of responding to requests for information			X	
Percentage of prepared "key messages" issued that appear in media coverage				X
Positive statements about the company made by regulators in public forums				X
Reduction in the number of activist group complaints/demonstrations made against the company				X
Cost incurred for communication/outreach training				X

PROCESS KNOWLEDGE MANAGEMENT	Catalogue Process Knowledge in a Manner that Facilitates Retrieval	Protect & Update Process Knowledge	Use the Process Knowledge	Maintain a Dependable Practice
The number of corrections to piping and instrumentation diagrams (P&IDs) and other process safety information identified during process hazard analyses (PHAs)				X
Number of PHA team recommendations that include an indication of less than adequate process knowledge where the information actually was available	X			
Accuracy of process knowledge during periodic reviews		X		
Backlog of change requests related to completing updates to process knowledge		X		
Results of random checks of process knowledge files after change requests are closed		X		
Number of PHA team recommendations that include an indication of less than adequate process knowledge where the information was not available				X
Percentage of document changes meeting the target cycle time for update				
Number of instances in which maintenance planners or purchasing agents cannot locate specifications or similar data	X			
Results of periodic opinion surveys to determine if users of process knowledge believe that it is current and accurate		X		
Results of random checks of material safety data sheet (MSDS) files to determine if they are complete, current, and accurate		X		
Frequency that process knowledge is accessed			X	

PROCESS KNOWLEDGE MANAGEMENT	Catalogue Process Knowledge in a Manner that Facilitates Retrieval	Protect & Update Process Knowledge	Use the Process Knowledge	Maintain a Dependable Practice
Number of incident investigations that include an element of discovery				X
Number of periodic reviews of information completed vs. plan				X
Average number of days required to have a drawing revised		X		
Number or percent of blank records in the process knowledge database				X
Results of periodic surveys to determine if users of process knowledge believe it is accessible	X			
If process knowledge is web-based, the number or percentage of dead links	X			
Number of times during audits or assessments that process knowledge (or duplicate copies of reports, etc.) must be retrieved from personal files		X		
Number of change requests initiated to "correct" process knowledge		X		
Engineering staff time spent recreating process knowledge		X		
Ratio of approved change requests (involving equipment changes) to updates to P&IDs		X		

HAZARD IDENTIFICATION & RISK ANALYSIS	Identify Hazards & Evaluate Risks	Assess Risks & Make Risk-based Decisions	Follow Through on Assessment Results	Maintain a Dependable Practice
Number of recommendations unresolved by their due date			X	
Number of hazard identification and risk analyses (HIRAs) of each type that are overdue				X
Percentage of repeat recommendations			X	
Number of incidents with risk analysis/assessment as a root cause of the incident		X		
Percentage of recommendations for administrative controls, active engineered controls, passive engineered controls, and inherently safer alternatives		X		
Number of recommendations per study or per year		X		
Average time corrective actions require for completion			X	
Percentage of recommendations rejected by management			X	
Management exceptions to risk criteria (accepting higher risk)			X	
Number of HIRAs of each type scheduled	X			
Technique used	X			
Ratio of actual losses to risk tolerance criteria		X		
Number of recommendations per revalidation		X		
Percentage of risk assessment teams meeting qualification/experience requirements				X
Average number or hours per P&ID for conducting baseline PHAs	X			
Average number of hours per P&ID for conducting PHA revalidations	X			
Time required to issue a HIRA report	X			

HAZARD IDENTIFICATION & RISK ANALYSIS	Identify Hazards & Evaluate Risks	Assess Risks & Make Risk-based Decisions	Follow Through on Assessment Results	Maintain a Dependable Practice
Percentage of intended revalidations that require the study to be completely redone				X
Number of qualified HIRA leaders, scribes, and participants				X
Percentage of new project risk assessments completed vs. plan				X

OPERATING PROCEDURES	Identify What Operating Procedures Are Needed	Develop Procedures	Use Procedures to Improve Human Performance	Maintain a Dependable Practice
Number of PHA recommendations related to inadequate operating procedures	X			
Number of standard operating procedures updated per year or staff hours spent updating procedures (per year, quarter, or month, depending on the review cycle)				X
Percentage of procedures revalidated per schedule/plan/period				X
Number of incident/deficiency reports related to procedures that were unclear, not available, or not widely understood			X	
Number of incident investigations that recommend changes to procedures			X	
Number of audit or assessment findings related to procedures missing some element of required content		X		
Number of units that have completed a task analysis to identify procedure needs	X			
If task analyses are periodically updated or revalidated, compliance with update/ revalidation schedule	X			
Number of special operating procedures developed to formalize the methods for completing infrequent or unusual tasks		X		
Number of management of change authorizations issued for each unit (per year) to permit temporary operations that recur on a periodic basis; in other words, issuing a temporary management of change authorization rather than a temporary procedure		X		
Percentage of procedures written by trained procedure writers				X
Number or operating personnel involved with reviewing and updating operating procedures				X

OPERATING PROCEDURES	Identify What Operating Procedures Are Needed	Develop Procedures	Use Procedures to Improve Human Performance	Maintain a Dependable Practice
Staff hours spent reviewing and approving procedures				X
Number of trained procedure writers				X

SAFE WORK PRACTICES	Effectively Control Non-routine Work Activities	Maintain a Dependable Practice
Percentage of permits completed correctly		X
Number of injuries related to nonroutine work	X	
Unsafe conditions or permit violations observed during routine audits (on a consistent basis)	X	
Number of nonroutine tasks that required a on-the-spot job safety analysis because there was no preexisting safe work procedure	X	
Number of near-miss incidents related to nonroutine work	X	
Number (or percentage) of safe work procedures that are past due for periodic review	X	
Number of causal factors identified by incident investigation teams related to failures to properly apply or follow a safe work permit	X	
Percentage of scheduled job observations/audits performed	X	
Compliance with the training plan for activities related to the safe work element		X
Number of management audits (per week or per month) of nonroutine work requiring permits		X
Frequency of improper shift-to-shift handoff.	X	
Number (or percentage) of safe work procedures revised each year	X	
Staff hours spent writing, reviewing, and approving safe work procedures	X	
Average time between a request for a permit and when it is issued	X	
Average time spent issuing a work permit, for example, total time spent issuing permits divided by the number of permits issued	X	
Progress toward implementing a new safe work practice or making significant improvements to the existing system		X

ASSET INTEGRITY & RELIABILITY	Identify Equipment and Systems that Are Within the Scope of the Asset Integrity Program and Assign Inspection, Test & Preventive Maintenance (ITPM) Tasks	Develop and Maintain Knowledge, Skills, Procedures, and Tools	Ensure Continued Fitness for Purpose	Address Equipment Failures and Deficiencies	Analyze Data	Maintain a Dependable Practice
Number (or percent) of overdue ITPM tasks			X			
Number of emergency/ unplanned repair work orders per month			X			
Percentage of ITPM tasks for safety-critical equipment that were completed on time						X
Number of temporary repairs currently in service (deferred maintenance items)				X		
Equipment reliability (or availability)					X	
Number of incidents with asset integrity/reliability as a root cause						X
Percentage of ITPM tasks that identify the need for immediate repairs, prior to putting the equipment or system back into service				X		
Number of equipment items included in the asset integrity program	X					
Total number of deferred repairs, such as known deficiencies that will be addressed at the next turnaround				X		

ASSET INTEGRITY & RELIABILITY	Identify Equipment and Systems that Are Within the Scope of the Asset Integrity Program and Assign Inspection, Test & Preventive Maintenance (ITPM) Tasks	Develop and Maintain Knowledge, Skills, Procedures, and Tools	Ensure Continued Fitness for Purpose	Address Equipment Failures and Deficiencies	Analyze Data	Maintain a Dependable Practice
Work order backlog for the inspection group, in other words, planned activities that are not yet past due			X			
Average time to address/correct deficiencies				X		
Number (or percentage) of ITPM tasks that uncover a failure					X	
Percentage of ITPM tasks managed in a risk-based fashion, rather than a simple periodic PM program					X	
Number of ITPM work orders (per month or quarter) that apply to equipment that is no longer present at the facility	X					
Number of inspectors/maintenance employees holding each type of required certification		X				
Total time charged to ITPM tasks each month/quarter			X			
Display a simple chart showing which facilities or units have fully implemented specific programs or practices						X

CONTRACTOR MANAGEMENT	Conduct Element Work Activities	Monitor the Contractor Management System for Effectiveness	Maintain a Dependable Practice
Percentage of incidents and near misses investigated by the facility that had root causes related to contractor activities		X	
Safety performance metrics for contractor companies		X	
Percentage of required contractor training conducted on schedule	X		
Number or frequency of negative findings in job safety evaluations, field inspections, audits of safe work practice implementation, and other safety-related audits		X	
Frequency of, and percentage attendance for, contractor safety meetings	X		
Percentage of contractor-related incidents or near misses that were subjected to a root-cause analysis	X		
Number of safety program improvement suggestions contributed by contractor personnel	X		
Frequency of required waivers of qualification requirements for past safety performance or current safety program			X
Percentage of qualification audits completed/qualification criteria met prior to entry		X	
Number of open contractor safety suggestions, in other words, those not yet resolved by a company representative, or average age for unresolved suggestions	X		
Percentage of contracted firms that, based upon post-job evaluation, would be considered for future contracts			X
Relevant statistics monitoring compliance with safe work practice procedures for contractor-involved jobs		X	

TRAINING & PERFORMANCE ASSURANCE	Identify What Training Is Needed	Provide Effective Training	Monitor Worker Performance	Maintain a Dependable Practice
Percentage of workers whose training is overdue	X			
Percentage of incidents with training and performance root causes				X
Time spent in training for individuals, shifts, departments, and job functions		X		
Percentage of workers who believe training is appropriate and effective		X		
Number of workers of each type whose training is overdue	X			
Number of qualified personnel in defined process safety management roles				X
Average test scores for classes, individuals, shifts, departments, and job functions		X		
Percentage of workers who require remedial training			X	
Percentage of workers who miss a particular test question			X	
Number of exceptions to training requirements				X
Percentage of correct answers during periodic unannounced tests administered to people in key process safety roles (line operations and management)			X	
Percentage of individuals who successfully completed a process safety management (PSM) training session on the first try			X	
Time spent on computer-based training (CBT) modules		X		
Number of errors during simulator training			X	
Percentage change in the training budget				X
Percentage of workers who miss a scheduled training session	X			
Percentage of training sessions that are offered on schedule	X			
Number of training sessions of each type scheduled	X			
Percentage of workers who test out of a training module			X	

TRAINING & PERFORMANCE ASSURANCE	Identify What Training Is Needed	Provide Effective Training	Monitor Worker Performance	Maintain a Dependable Practice
Number of subject matter experts providing training				X
Percentage of course deliveries audited				X

MANAGEMENT OF CHANGE	Identify Potential Change Situations	Evaluate Possible Impacts	Decide Whether to Allow the Change	Complete Follow-up Activities	Maintain a Dependable Practice
Number of incidents with management of change (MOC) as a root cause					X
Percentage of work orders/requests that were misclassified as replacement-in-kind (RIK) (or were not classified) and were really changes	X				
Percentage of temporary MOCs for which the temporary conditions were not corrected/restored to the original state at the deadline				X	
Percentage of MOCs reviewed that were in full compliance with the site's MOC procedure					X
Number of "emergency MOCs" or the ratio of "emergency MOCs" to total MOCs					X
Percentage of MOCs for which the drawings or procedures were not updated				X	
Percentage of changes that were reviewed within the MOC system but were reviewed incorrectly		X			
Percentage of MOCs reviewed that were not documented properly				X	
Number of MOCs performed each month					X
Percentage of MOCs reviewed that did not have adequate hazard/risk analysis completed					X
Percentage of MOCs for which the workers were not informed or trained				X	
Percentage or variation in the number of changes processed on an emergency basis					X

MANAGEMENT OF CHANGE	Identify Potential Change Situations	Evaluate Possible Impacts	Decide Whether to Allow the Change	Complete Follow-up Activities	Maintain a Dependable Practice
Average backlog of MOCs/active MOCs					X
Ratio of identified undocumented changes to the number of changes processed by the MOC program	X				
Percentage of recent changes that involved the use of backup MOC personnel		X			
Percentage of changes that were properly evaluated, but did not have all authorization signatures on the change control document			X		
Monthly average in the percentage of work requests classified as a change					X
Percentage of personnel involved in the MOC system who believe the system is effective					X
Difference between the percentages of senior managers and routine users who believe the MOC program is effective					X
Average amount of calendar time taken between MOC origination and authorization					X
Average number of staff-hours per MOC from the time the MOC is originated until the time the MOC is approved for implementation					X

OPERATIONAL READINESS	Conduct Appropriate Readiness Reviews as Needed	Make Startup Decisions Based upon Readiness Results	Follow Through on Decisions, Actions, and Use of Readiness Results	Maintain a Dependable Practice
Number of startups for which readiness reviews were not performed	X			
Number of action items overdue			X	
Number of incidents that occur during startup				X
Number of readiness reviews for which authorizations to restart were not found		X		
Number of spurious shutdowns after startup				X
Number of readiness reviews performed	X			
Number of startups deferred as a result of problems found during readiness reviews		X		
Number of personnel trained prior to startup				X
Number of issues during startup that should have been discovered during the readiness review			X	
Number of improperly assembled pieces of equipment found during readiness reviews				X
Staff-hours expended on readiness reviews				X
Time from readiness review to completion of all action items			X	
Duration of startup				X
Amount of off-spec product or loss of raw material as a result of startup problems				X
Number of people trained per year on readiness				X

CONDUCT OF OPERATIONS	Control Operations Activities	Control the Status of Systems and Equipment	Develop Required Skills/ Behaviors	Monitor Organizational Performance	Maintain a Dependable Practice
Number of nonroutine and emergency maintenance work orders		X			
Number of nuisance and always-on alarms		X			
Number of incidents during which safe operating limits were exceeded	X				
Percentage of overtime hours			X		
Number of requests made to bypass critical interlocks					
Incidence of shortcuts identified by near misses and incidents	X				
Number of housekeeping audits and their scores		X			
Number of unplanned shutdowns				X	
Number of unplanned safety system activations for valid reasons				X	
Number of incidents with operations issues as root causes					X
Number of incomplete shift logs or reports		X			
Number of times workers are challenged to solve "what-if" scenarios			X		
Number of human-machine deficiency and near miss reports	X				
Number of incidents attributed to trainees	X				
Number of missed surveillance rounds		X			
Number of incidents caused by a lack of self-checking or peer-checking			X		

CONDUCT OF OPERATIONS	Control Operations Activities	Control the Status of Systems and Equipment	Develop Required Skills/ Behaviors	Monitor Organizational Performance	Maintain a Dependable Practice
Number of manager inspections of work locations				X	
Percentage of workers failing random substance abuse tests				X	
Staff turnover rates					X
Number of distributed control system (DCS) points that have been taken off scan		X			
Number of visitors to the control room	X				
Number of audit findings related to inoperable instruments and tools		X			
Average time to complete repairs on the human-machine interface		X			
Average time to resolve off-normal findings		X			
Number of access permits issued		X			
Absenteeism			X		
Average time required to complete required reading			X		
Frequency of communication of progress toward goals				X	
Percentage of manager inspections delegated to subordinates				X	
Number of disciplinary actions				X	
Number of qualified personnel in defined roles					X
Progress toward performance goals					X

EMERGENCY MANAGEMENT	Prepare for Emergencies	Periodically Test the Adequacy of Plans and the Level of Preparedness	Maintain a Dependable Practice
Number of trained environmental response team (ERT) members on each shift	X		
Percentage of failed tests or inspections of emergency response equipment	X		
Percentage of ERT members who are up-to-date on emergency responder training requirements		X	
Percentage of sites that conducted a drill with local emergency responders during the year		X	
Percentage of preventive maintenance work orders for emergency response equipment that are past due	X		
Fraction of drills that are conducted as scheduled		X	
Number of meetings or other contacts with local emergency responders or the local emergency planning committee (LEPC) regarding emergency response plans			X
Number (or percentage) of units that have up-to-date plans	X		
Number of errors/omissions in the emergency response plan or its annexes discovered during drills and training	X		
Number (or percentage) of emergency response plans or annexes to the plan that are past due for periodic review	X		
Results of opinion surveys among operators regarding their perception of the unit's or facility's state of preparedness for emergencies		X	
Number of meetings or other contacts with the community regarding how they will be notified of an emergency and what they should if they are notified			X
Supply status for ERT consumable supplies	X		
Number of changes to the emergency response tactics or logistics based on critiques of drills or other exercises		X	

MEASUREMENT & METRICS	Use Metrics to Make Element Corrective Action Decisions	Conduct Metrics Acquisition	Maintain a Dependable Practice
Number of metrics for which data are collected	X		
Frequency of communicating metrics	X		
Number of process or procedure changes resulting from learnings from metrics/data			X
Refresh rate for metrics	X		
Percentage of management personnel that use metrics for decision-making		X	
Frequency of metrics use in management review meetings		X	
Number of risk-based process safety (RBPS) elements for which metrics are maintained			X
Number of audit findings dealing with the metrics element			X
Number of metrics communications tools developed	X		
Percentage of employees who have seen metrics		X	
Number of problems avoided/discovered through the use of metrics		X	
Results of audits or management reviews indicating that metrics are in consistent use			X
Percentage of surveyed personnel who understand the relevance of specific metrics			X
Staff-hours required to develop metrics	X		
Number of people trained on metrics element			X

AUDITING	Use Audits to Enhance RBPS Effectiveness	Conduct Element Work Activities	Maintain a Dependable Practice
Percentage of audit findings that are repeat findings		X	
Average and maximum number of days overdue for open recommendations	X		
Number or percentage of unresolved audit recommendations	X		
Percentage of audits completed according to schedule			X
Percentage of audit recommendations resolved on time			X
Percentage of key management personnel that attended the audit closeout meeting			X
Percentage of near-miss and incident investigations identifying RBPS management system weaknesses that were not detected by prior audits			X
Number or percentage of auditors meeting qualification requirements			X
Number of person-days required to complete an audit	X		
Trends in the number or significance of findings over a series of audits of the same facility		X	
Percentage of recommendations that are rejected by the facility management		X	
Percentage of scheduled audits that are completed on time			X
Interval between completion of onsite work and completion of the audit report	X		
Number of previous audits conducted by each audit team member			X

MANAGEMENT REVIEW & CONTINUOUS IMPROVEMENT	Conduct Review Activities	Monitor Organizational Performance	Maintain a Dependable Practice
Number of management reviews per time period	X		
Number of repeat findings in reviews			X
Number of deficiencies identified by management reviews	X		
Type and number of findings in audits		X	
Number of incidents attributed to RBPS element failures		X	
Time required to resolve deficiencies identified by management reviews	X		
Percentage of reviews delegated to subordinates		X	
Changes in performance goals			X

INCIDENT REPORTING & METRICS	Process Safety Incidents	Process Safety Near Misses	Incident/ Near-Miss Subcategories and Cause Types
Process Safety Total Incident Rate (Total PS incidents x 200,000/total employee & contractor work hours)	X		
Number of Incidents (releases, fires, explosions) that meet reporting thresholds per year	X		
Number of near-miss events (as defined in Chapter 3)		X	
Process Safety Incident Severity Rate (Total severity score for all PS incidents x 200,000/total employee, contractor & subcontractor work hours)	X		
Process Safety Rate by Severity Levels (Total PS incidents at severity level "X" x 200,000/total employee, contractor & subcontractor work hours)	X		
Number of near-miss events (as defined in Chapter 3) with potential for significant consequences		X	
Reportable injuries per 200,000 employee hours worked	X		
Damages ($ at/above threshold) -- on-site	X		
Number of incidents per 200,000 work hours	X		
Reportable injuries per 200,000 contractor/subcontractor hours worked	X		
Emergency response/shelter in place called -- on-site	X		
Additional incident information can be collected and statistics created from the perspective of timeframe (day of week, time of day, month of year), the type of unit operation (manufacturing plant, supply chain, transportation), the state of operation (normal, startup, shutdown, other), or different cause types (faulty equipment design, corrosion, failure to follow procedure, etc.). This can create a number of additional metrics.			X
Number of incidents by severity	X		
Reportable injuries -- third party	X		

INCIDENT REPORTING & METRICS	Process Safety Incidents	Process Safety Near Misses	Incident/ Near-Miss Subcategories and Cause Types
Fatalities per 200,000 employees hours worked	X		
Fatalities per 200,000 contractor/subcontractor hours worked	X		
Fatalities -- third party	X		
Damages ($ at/above threshold) -- off-site	X		
Emergency response/shelter in place called -- off-site	X		
Number of near-miss events (as defined in Chapter 3) per 200,000 man-hours worked		X	
Number of near-miss events (as defined in Chapter 3) with potential for significant consequences per 200,000 man-hours worked		X	

APPENDIX II: PROCESS SAFETY PERFORMANCE INDICATORS: BP CHEMICALS HULL CASE STUDY

BACKGROUND

The Hull site produces organic acids and derivatives. As part of an initiative launched by the UK Health & Safety Executive the site developed a number of Process Safety Performance Indicators (PSPI) in accordance with the Health & Safety Executives guidance document HSG 254, which describes a process for the development of PSPI at an individual plant level. This document describes how that activity was undertaken and the subsequent learning's from the process together with the planned wider rollout.

INITIAL PROCESS OF REVIEW

In order to gain experience in the process outlined in HSG 254 it was agreed that a Hull site plant would be used in the pilot. A review session was planned with operational and technical representatives from the plant together with the local Health & Safety Executive Regulatory Inspectors and the author of HSG 254 who lead the review process.

The Hull site is a top-tier COMAH establishment and the individual plant safety reports were used as a base to identify the major hazard scenarios. While the Hull plants produce different final products, the key operational stages are the same, i.e., feed system, reactor section, initial separation and recycling, and final distillation train. The Risk Control Systems (RCS) were therefore fundamentally the same but each plant was reviewed in isolation. A total of eight different RCS were considered as part of the review and the process described in HSG 254 was used for each one. The Workbook produced during the review for one of the plants as part the process is attached in Appendix 1.

For each of the RCS two key questions were considered:

1. What are the most important parts of the RCS that must be present all the time?
2. What does successful implementation of the RCS look like?

The answers to the first question would provide leading indicators while answers to the second would provide lagging indicators. For the 8 RCS chosen, a total of 8 lagging and 18 leading indicators were identified. It was then necessary to carry out a "beauty contest" of the RCS and determine a single leading and lagging indicator for each RCS. This resulted in choosing 5 RCS and 10 indicators and these are presented in Table 1 below. The selection of RCS reflects the high level of automation of the plants.

Table 1 Agreed RCS and PSPIs

RCS	Lagging Indicator	Leading Indicator
Equipment Overpressure	Number of high-pressure alarms	Number of times the condition of all pressure control loops falls below set standard
Accidental Leakage	Number of high & medium priority leaks as defined in SSO 307	% completion of planned process technician routines
Equipment Overfill	Number of high-level alarms	Number of times the condition of all level control loops falls below set standard
Corrosion	Number of times corrosion rate exceeds predicted rate	Number of Ultrasonic Thickness (UT) checks not completed on time
Management of Change	Number of incidents with MOC as a root or system cause	Number of risk review comments added at Modification Approval Form (MAF) stage

These final PSPIs were agreed and actionees nominated for the collection and capture of the data.

PSPI GENERATION AND THEIR EFFECTIVENESS

For equipment overpressure and overfill, the lagging indicators were identified as the number of high-pressure or high-level alarms that occurred on the plant and was captured directly from the Honeywell Distributed Control System (DCS) system. These alarms are precursors to loss of containment scenarios which in themselves would lead to either a flash fire, vapor cloud explosion (VCE), or toxic release, and they thus identify events where actual operation has moved beyond the normal operating envelope. Rather than collecting alarm data for just the top-level events, such as failure of the reactor or flash drums, it was decided that the data would be collected for every high-pressure and high-level alarm configured on the plant. The lagging indicators for these RCS were chosen as the number of pressure and level control loops that operated outside their normal configuration (this is essentially when the control loop is in manual rather than auto or when it is auto local rather than cascade and also when the output of the controller is saturated either at 0 percent or 100 percent output—in these situations the DCS control is not effective and will not quickly return the process to within its operating envelope). Data collected for the two RCS is presented graphically in Table 2 (see next page).

These indicators have showed that there were more alarms occurring than we had expected and this was despite us having been tracking the Engineering Equipment & Materials Users Association (EEMUA) alarm metric of the number of alarms per operator per 10 minutes and reporting measurements of between 1.5 and 2.0. It was also surprising that the number was as high as it was when the control loop count (leading indicator) was as low as it was, which may indicate that the alarm settings may be set too tight.

Table 2 Overpressure and Overfill Indicators

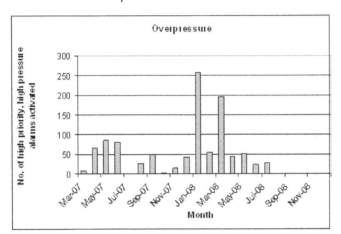

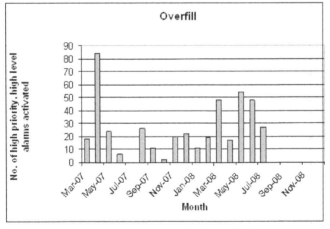

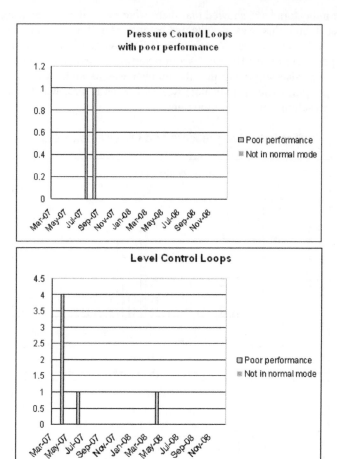

Releases of hazardous materials can also occur due to loss of containment from other damage mechanisms, e.g., corrosion, poor maintenance, and the indicators for accidental release RCS are presented in Table 3 (see next page):

Table 3 Accidental Releases

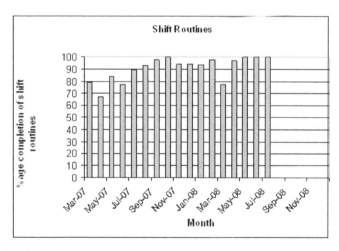

The lagging indicator was relatively easy to track and is simply the number of process leaks identified. There was no minimum threshold of quantity of release for this data and so it covers both small drips and larger releases. The leading indicator was more difficult to determine, but it was agreed that proactively operating technicians should be out checking the plant using their eyes, ears, and noses to ensure any accidental releases are quickly identified and any substandard conditions are addressed before they result in a release. The level of completion of operating technician routines saw a gradual improvement—"what gets measured gets done."

Due to the nature of the hazardous materials used in the processes, corrosion is a key mechanism for release of materials and so a separate RCS around corrosion was identified. The leading indicator was quickly identified as being the completion of all inspection activities to plan, and since this review started there have been no inspections missed. The suitability of an indicator that has consistently been at zero is clearly questionable and we will need to consider if there is a better indicator. The lagging indicator is around the output of the inspection process where there has been a reduction in the interval between inspections; i.e., it is a measure that the condition of the plant is not as expected. Again the data collected is presented in Table 4 below:

Table 4 – Corrosion

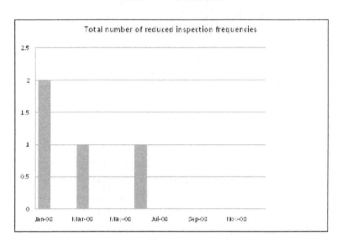

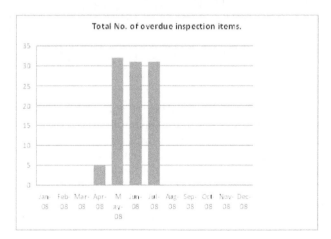

The final RCS chosen was that of management of change (MOC) as this is a key area where asset integrity can be affected by not executing MOC with the necessary rigor (see Table 5 on next page). The lagging indicator is set as the number of incidents where the execution of the MOC was a root cause. The leading indicator was set as the number of corrections or clarifications sought by the MOC approval signatories. (The MOC process requires a multi-disciplined team to review the proposed change, which should identify all the actions required to execute the MOC safely. Once this review has been completed and the details around the change have been agreed, it is then forwarded for approval prior to execution.) This indicator tracks the number of corrections and clarifications made by the senior approving signatory and is thus a measure of the quality and integrity of the MOC system. The leading indicator has shown a marked improvement in the quality of MOC and again has highlighted the benefit of measurement and reporting as part of demonstrating safety leadership expectations.

Table 5   Management of Change

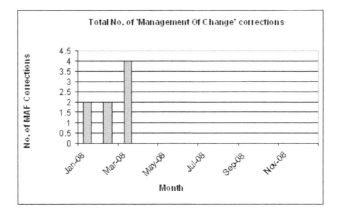

OVERALL VIEW ON THE PROCESS AND THE WAY FORWARD

Overall the process has identified a number of areas where further action is necessary (alarm reviews) to ensure clarity and integrity of operations to reduce still further the risk of a major accident. There are also a number of metrics that have not worked as well as expected (overdue inspections and operating routine execution) and will require further work to identify better measures. The process as defined in HSG 254 recognizes that this is an ongoing process and indicators that show little change should be reviewed and others tracked instead.

One of the requirements rolled up from plant level, through a site and up to the organizational level. Of the metrics identified in this review most of them could be captured at a site level and would be valid for other chemical plants at the Hull site. Indeed the indicators for overpressure and overfill are already been tracked at a site level. However, rolling these metrics up to an organizational level in a major global company with diverse operations like BP would prove difficult and not all of them would allow a reporting hierarchy to be developed. That said, the process of reviewing the individual plant assets and thinking through what the RCS are and the key elements of their delivery is in itself a beneficial exercise and allows the spotlight to be placed on the key issues of process safety.

APPENDIX III: NOVA CHEMICALS UNCONTROLLED PROCESS FIRE AND LOPC METRICS

In 1996, Jeff Lipton, CEO of NOVA Chemicals, set a goal for NOVA Chemicals to reduce the number of uncontrolled process fires to zero as part of an effort to reduce the risk of a catastrophic process incident.

Process fire targets are routinely reviewed and made part of everyone's objectives, including Mr. Lipton's. The Executive Leadership Team and Mr. Lipton are notified of each fire, regardless of size, damage, or impact. NOVA Chemicals missed ambitious fire reduction targets initially, but began to improve at faster and faster rates. It became obvious that if materials are kept from leaking out of pipes, pumps, valves, process vessels, and storage tanks, process fires, even in the most dangerous operation, are greatly reduced or eliminated. NOVA Chemicals found measures that are leading indicators for successful fire safety—the key one being loss of process containment (LOPC). NOVA Chemicals quickly developed leading LOPC metrics and began monitoring these alongside uncontrolled process fire data.

Since the initial use of these metrics, NOVA Chemicals has reduced uncontrolled process fires from 65 in 1998 to 6 in 2007. The uncontrolled process fire target for 2008 is 5 or less.

NOVA Chemicals believes that the next big safety step the industry can take is to measure and publicly report loss of process containment and uncontrolled process fires.

BACKGROUND INFORMATION

Definitions

- Process—An area where process materials are manufactured, stored, handled, or otherwise used, including all utilities and electrical and ancillary equipment associated with these areas. Similar functional areas in pilot plants and laboratories fall under this definition.
- Fire—An unintended oxidation occurring in a process area that produces flame or glowing embers, or evidence that this has occurred, such as charred or burned material.
- Controlled Process Fire—The fire potential was anticipated and safeguards put in place to control/contain the fire should it occur, *and* the fire did not exceed the anticipated consequences, *and* there was no

damage to equipment beyond the initiating failure, *and* there were no injuries to personnel resulting directly from the fire.

- Uncontrolled Process Fire—Any process fire that cannot be classified as controlled.

- Loss of Process Containment (LOPC)—An incident that involved an unanticipated leak, spill, or release of process material in sufficient quantity or concentration to the air, water, land, or work environment that resulted or could reasonably have resulted in a process, safety, or environmental incident.

Use of NOVA Chemicals' Process Fire Definition

Once it is determined there was a fire in the process area, the task becomes determining whether it was controlled or uncontrolled. By defining a controlled fire only, it eliminates any of the gray area between the two definitions. Essentially, if it does not fit the definition of controlled, it is uncontrolled.

The definition of a controlled process fires needs to be broad enough to encourage personnel to put in place safeguards to prevent a fire from becoming uncontrolled. To do this, the fire must be anticipated and planned for as part of work activities. This means all work activities including ongoing operations. In this way, fired process heaters, pyrophorics, hot work, electrical fires inside a fire-resistive enclosure, hydrogen PSVs that ignite when vented to a safe location, among others, are all covered. The key is to identify that the fire could occur and then put in place the appropriate safeguards.

The other aspects of the definition state that if there is a fire and it is anticipated, the damage cannot be greater than the failure that was anticipated. If a pyrophoric fire is anticipated and safeguards are put in, but the fire burns down a unit, it was obviously not controlled since there was damage beyond the initial failure. The same is the case for injury to personnel as a direct result of the fire. (This does not include someone spraining an ankle while retrieving a fire extinguisher, for example).

Application of this definition should not only be used after a fire has occurred; rather it can be used before the fact. During the design phase, this definition can be used to identify fire hazards. This allows the engineer to design means to control process fires should they occur. For example, when designing PSVs, a decision needs to be made as to where it should discharge. By discharging hydrogen vents in a safe location, any ignition would be deemed a controlled fire, provided no further damage occurs.

The definition can also be used during pre-job safety reviews. Hot work is an obvious example. A proper procedure to handle pyrophoric materials is another example.

INDEX

☼ *indicates material on ftp site*

☼ indicates material on ftp site

☼ *indicates material on ftp site*

☼ indicates material on ftp site

☼ indicates material on ftp site

Printed and bound by CPI Group (UK) Ltd, Croydon, CR0 4YY

16/04/2025

14658584-0005